1998

D0049036

LIFE SIGNS

THE BIOLOGY OF STAR TREK

LIFE SIGNS

THE BIOLOGY OF STAR TREK

**SUSAN JENKINS, M.D., and
ROBERT JENKINS, M.D., PH.D.**

Foreword by Lawrence M. Krauss, Ph.D.,
author of *The Physics of Star Trek*

HarperCollins*Publishers*

LIFE SIGNS—THE BIOLOGY OF STAR TREK. Copyright © 1998 by Robert Jenkins and Susan Jenkins. All rights reserved. Printed in the United States of America. No part of this book may be used or reproduced in any manner whatsoever without written permission except in the case of brief quotations embodied in critical articles and reviews. For information address HarperCollins Publishers, Inc., 10 East 53rd Street, New York, NY 10022.

HarperCollins books may be purchased for educational, business, or sales promotional use. For information please write: Special Markets Department, HarperCollins Publishers, Inc., 10 East 53rd Street, New York, NY 10022.

FIRST EDITION

Designed by Elliott Beard

ISBN 0-06-019154-6

98 99 00 01 02 ❖ / RRD 10 9 8 7 6 5 4 3 2 1

For Kathy and Chris
and in memory of Carl Sagan
and Gene Roddenberry

CONTENTS

FOREWORD

One of the first "fan" letters I received after writing *The Physics of Star Trek* was from someone who said he had been waiting for thirty years to read a *Star Trek* book in the science FACT section of a bookstore! I hadn't fully appreciated, even when writing the book, the degree to which *Star Trek* had permeated our culture, and how much it inspires people to imagine the "untold possibilities of existence," as the supernatural prankster, Q, once said.

One of *Star Trek*'s chief attractions is its focus on the dramatic possibilities presented by extraterrestrial civilizations and alien psychologies. This, of course, creates numerous challenges to anyone intent on reconciling the biological principles of the series with those of our own world. We might wonder for instance, why there are so many humanoid life-forms on *Star Trek* (even while we realize that most Hollywood extras come with one head, two arms, and two legs as standard equipment). It's also hard to imagine how a hybrid like Spock could arise from the mating of a human and a Vulcan when interspecies mating on Earth hardly ever succeeds, even when the species share over 90 percent of the same genetic building blocks, not to mention an evolutionary path confined to the same planet. Of course, our popular culture would have suffered if Spock had never been born, and Gene Roddenberry knew just how and when to bend the rules of science so that *Star Trek*'s drama would be more vital and compelling.

Even while we indulge the biological liberties *Star Trek* takes with the viability of various exotic and not-so-exotic life-forms, we should remember the string of recent discoveries that have called into question some of our own most cherished notions about the nature and origin of life on Earth. Life seems to exist on Earth, for example, in all the wrong places. In toxic environments, scalding water, and barren tundra, so-called "extremophiles" have evolved and survived. It's even possible that the first forms of life on Earth arose in this context.

Who are we, in fact, to say that life on Earth even began on Earth! Recent observations of the Hale-Bopp comet provide evidence of complex organic molecules—the basis of amino acids—in space. Could life on Earth have evolved so rapidly after the physical conditions calmed down enough for it to survive, because the fundamental constituents had been created earlier and elsewhere?

Compelling questions like these strongly suggest that the science of biology is now undergoing the sort of excitement that attended the revolutions of early-twentieth-century physics. This is a thrilling time in the life sciences, and if *Star Trek* can excite people to ponder the wonders of nature and the possibilities of life, then it can serve an important and serious purpose in addition to being fun. As this book amply proves, examining *Star Trek*'s universe can reveal many fascinating hidden truths about our own. And vice versa.

LAWRENCE M. KRAUSS
CLEVELAND, OHIO
FEBRUARY 1998

PREFACE

This book owes its existence to Dr. Lawrence M. Krauss of Case Western Reserve University and his hugely successful 1995 book *The Physics of Star Trek*. We read Dr. Krauss's book and enjoyed it immensely. The natural next step was to voyage through the *Star Trek* centuries ourselves, making use of our professional backgrounds in psychiatry and genetics.

We proceeded to gather specimens of alien species and extraterrestrial humanoids throughout the *Star Trek* universe, and in this task the crews of the various generations were quite helpful: Spock contributed a number of trenchant scientific observations, Dr. Beverly Crusher was crucial, Data had a lot to say, Kathryn Janeway managed to avoid making too much of a mess of things, and Quark negotiated us through some tight spots. Having collected our specimens, we returned to Earth to examine them in the cold light of reality. A lot of them didn't survive the trip.

But some of them did! And that is the fascination of Gene Roddenberry's great work of art. *Star Trek* is a phenomenon like few others, lending itself to almost limitless elaboration in many fields of human endeavor. By projecting ourselves "out there," we human beings become more aware of who we are. Through nearly thirty years of production—and thanks to countless writers, producers, special-effects artists, actors, commentators, and even spoofers who have contributed to this grand adventure—*Star Trek* reflects how we are changing as well. Just to take a few examples: consider the great difference between the roles and costuming of women in the original series and in the *Voyager* series; Benjamin Sisko as commander of Deep Space 9; Jean-Luc Picard's disciplined form of diplomacy as opposed to James T. Kirk's bellicosity and occasional recklessness. There is much that we need to learn in order to survive as a species, and *Star Trek* can be a vehicle for that exploration. Professor Eric Rabkin of the University of Michigan refers to science fiction as "the literature of the technical imagination." According to Rabkin, a work of science fiction requires three elements: the fantastic made plausible, high drama, and intellectual excitement. It is the intellectual excitement that grabs scientists who watch *Star Trek*. If this book stirs up a little intellectual excitement about real science for some *Star Trek* fans, we are well pleased.

From our colleagues in the sciences, we ask your patience and indulgence. We found it very difficult, while writing this book, to steer a course through the complex gravitational field generated by the binary star system of Scientific Accuracy and Brevity in Plain English. Although this book is primarily about *Star Trek*, we have tried to touch on many of the fascinating research questions currently challenging biologists. Scientists are an honorable tribe who share with the Vulcans a dedication to honesty and a painstaking love of detail. We hope we have accurately communicated the core scientific concepts. If we have occasionally triggered a factual meltdown, we apologize and ask that you will let us know.

Our other task was to select from all the universe a few real lifeforms to examine as illustrations. In some cases we chose the common and familiar; in others we focus on the extraordinary. This is how biological understanding is built. Through encounters with the fascinating oddity we can better recognize the prevailing general principles.

This book would not have been written without the sponsorship of our agent, Susan Rabiner, who also contributed invaluable editing assistance. Eamon Dolan of HarperCollins never yielded on deadlines or flagged in his positive outlook. Sara Lippincott provided hours of editorial expertise. We also wish to thank a number of friends and colleagues, particularly Stephen Carmichael, Rhonda Erdman, Anna Fortunato, Chris Hook, Robert Johnson, Doug Nichols, Ron Reeves, Carol and Steve Solovitz, Jane Toft, and David Wold, for their encouragement and suggestions.

We also thank our parents, who, during their visits to our home and our visits to theirs, put up with years of viewing and discussing science fiction television shows and books, for which they have little appetite. Last but not least, this book would not have happened without the unfailing interest, love, and support of our two children, Chris and Kathy. We enjoyed debating the fine points of *Star Trek* ("Dad! There are at least *three* kinds of hand-held phasers") at the dinner table, in the car, and in front of the TV. We're not sure which of our two generations enjoyed watching all the reruns and new episodes more!

Sometimes, if you follow your bliss, it takes you somewhere.

ROCHESTER, MINNESOTA
FEBRUARY 1998

LIFE SIGNS

THE BIOLOGY OF STAR TREK

CHAPTER ONE

A Face Only a Mother Could Love

"It's amazing you all have gotten this far, you all have such featureless foreheads."

—*Seska to the* Voyager *crew ("Basics, Part I")*

*T*he Place: Camp Khitomer, near the Klingon-Romulan border. *The Scene: A hall in which ambassadors from all parts of the United Federation of Planets are gathered to consider an emergency application from the Klingon homeworld and its colonies for admission to the Federation. Suddenly, Kirk, Spock, Scotty, McCoy, Chekov, and Uhura burst through the security barriers. While McCoy and Uhura push through the crowd, Kirk leaps across the podium and knocks the president of the Federation to the floor. Whooosh!! Phaser fire sears the air. A glass ceiling panel shatters, and a masked would-be assassin falls to the floor. Dead.*

It's over before anyone realizes the danger. Confusion and terror give way to relief. The delegates burst into applause. Shaken but unharmed, the Federation president is helped to his feet, smooths his white beard, and tries to compose both his attire and his dignity. Our heroes—Kirk and the crew of the Enterprise—*all smiles and modesty, accept his congratulations. Good has triumphed over evil. The Federation is safe again—for the moment.*

And so, more or less, concludes the sixth *Star Trek* movie, *The Undiscovered Country.* It is late in the twenty-third century. Jean-Luc Picard has not yet been born, and only part of the galaxy's Alpha Quadrant has been explored. And yet, as the camera pans around the conference room and we see the gathered delegates, now relaxed and smiling, their faces, while somewhat familiar to a devoted Trekker like you, are nevertheless distinctly . . . not like ours. Not . . . *human.* There are blue-skinned Andorians, and we recognize many Klingons and a few snouted Tellarites, but where on earth (or should we say "off earth"?) are those web-headed people in the front row from?

THE MAKING OF AN ALIEN

The *Star Trek* episodes from the original series *(TOS),* with a few exceptions, featured human-looking aliens who were merely colorful, like the green-skinned Orions. But scientifically speaking, is there any reason that aliens should look so much like us? Not really—unless, of course, we all evolved from a common ancestor. The aliens in the subsequent series—*The Next Generation, Deep Space Nine,* and *Voyager*—become increasingly out-of-this-world, with facial appendages, extra orifices, feathers, spots, dots, horns, and skin tags flapping and waving from every surface. Fortunately for the sake of human actors, the basic bipedal, upright humanoid form is preserved throughout most of the galaxy. The *Star Trek* writers attempted to account for this phenomenon in *The Next Generation (TNG)* episode "The Chase." Archaeological research by Professor Richard Galen, we learned, led to the discovery of an ancient humanoid race that had seeded its DNA throughout the galaxy, giving rise to terrestrial humans, Vulcans, Romulans, Klin-

gons, Cardassians, and other humanoid races. This idea approxi-
mates the Panspermia hypothesis of the nineteenth-century physicist Svante Arrhenius, who proposed that life got started on Earth
from primitive spores that drifted here from outer space. His notion
was elaborated in the 1970s by biologists Francis Crick (the codis-
coverer, with James Watson, of the molecular structure of DNA,
the genetic material) and Leslie Orgel, who suggested that Earth's
lifetime of 4.5 billion years had not been long enough for life as we
know it to have arisen unaided, and that therefore the planet had to
have been seeded with microorganisms, perhaps dispatched here
deliberately, in the nose cone of an alien rocket ship. If this turns
out to be the case, we might one day discover life throughout the
galaxy on other likely planets. The Panspermia hypothesis cannot
begin to be tested unless or until extraterrestrial life is discovered.

But, you say, recalling your biology courses, if Galen was right,
wouldn't enough time have passed after all those generations for
Star Trek's alien species to have grown farther apart? By now,
shouldn't they all look vastly different from one another? Do they
all look so much alike because the *Star Trek* makeup artists had to
work with human actors? Was it paucity of imagination that limited
the forms of *Star Trek's* aliens, or biological faithfulness? What con-
straints on rapid and dramatic speciation does evolution impose?

The evolutionary clock runs slowly, and it has a built-in bias
against major overhauls. Because speciation is brought about by
multiple random changes in DNA and the changes must allow the
organism to survive and reproduce, small changes are favored over
large ones. Small changes are less likely to compromise the tested
survivability of the original. Given this constraint, two species that
start out alike can remain similar over a rather long time, even
under very different sets of environmental pressures.

You can get a better feel for what's at stake here by playing a
child's game. Choose any two four-letter words with no letters in
common, such as "boat" and "like." Try to change the first word

into the second by changing only one letter at a time. What makes the game a game is that each change must leave you with four letters that spell a word, and the shortest route wins. (We came up with boat-bolt-bole-bile-bike-like: five steps, and maybe you can do better.) This is, more or less, how our genes work. The individual letters of our DNA can change, but after each change, our DNA must still represent a program for a viable person. If two groups of humanoids start with the same "word" but end up at different target "words," the intermediate words in both groups will not be all that different from each other, simply because the intermediate words must all be words themselves. Something similar happens with evolution and DNA.

Does it follow that a *Star Trek* makeup artist can make a change in any feature of an Earth-type face and come up with a viable variation as long as the change is kept small? Alas, not entirely. Other factors affect which traits can emerge, still others select which traits will survive, and still others determine which traits will be passed on to future generations.

Building on Dr. Galen's research, let us suppose that some primordial humanoid seeded common DNA throughout the galaxy—DNA that coded for similar basic body plans. Those bodies would still evolve to deal optimally with their own special environmental pressures. Our bodies, for instance, are adapted to the conditions of life on planet Earth, where the atmospheric pressure at sea level is 10,332 kilograms of force per square meter (15 pounds per square inch), with gravity at 1 G-force and surface winds and temperatures limited to a certain range. Evolution in other environments would encourage retention of spontaneous mutations, however small, that increased the fitness of the humanoid species in those environments. Thus, for example, because Vulcan is a desert world with little cloud cover to block out the rays of its sun, Vulcans evolved with a second, internal eyelid that acts like human sunglasses. Harsh conditions, particularly the greater grav-

ity (Vulcan is more massive than Earth) led to increased musculature, giving Vulcans a superior strength and agility. The Trill and
the Bajorans, on the other hand, come from homeworlds that
more closely resemble Earth, and they more like Terrans (that is,
terrestrials, or Earthlings, like us). Only traits that give the
humanoid individuals selective survival advantage will spread
throughout the species and contribute to the evolutionary pattern
of the race.

To inject a little verisimilitude into the design of *Star Trek* aliens
(and to form some idea of what might be out there, should
humanoid life exist somewhere else in the universe) we have to
abide by some basic biological rules. Only then can we be sure
which *Star Trek* aliens could be for real and which come straight
out of central casting.

HOW HUMAN FACES ARE MADE

One of the basic rules is that development of organisms begins
from a single cell, whether you are an Earthman, a Klingon, or a
Whoever. This single cell results from the fertilization of an egg
cell by a sperm cell and the consequent combination of the
parental genes. The fertilized cell divides to become a ball of cells
known as a morula, but it hasn't grown yet; it is still the same size
as the original fertilized egg, but now it has been sectioned up into
a small number of cells—from sixteen to sixty-four of them. Now
comes the really terrific part.

Each cell has to start turning itself into a specialized part of the
organism. Eventually, every cell in the organism will have a job to
do, and somehow all the jobs must be assigned so that no jobs are
left undone. This dividing up of assignments is called cell differentiation, and it is a complex process. But it starts right here in the
morula: the cells of the morula already "know" which of its two

poles they are on, and they have begun to specialize accordingly. The RNA from the egg cytoplasm—RNA (ribonucleic acid) being the version of the genetic material that conveys the DNA instructions to the protein factories of the cell—is the first instigator, or switch, in a cascade of changes that will lead to increasing specialization among the cells of the developing embryo. But here's the kicker: if something goes wrong—even a small something, like the loss of a bit of chromosome in one cell of the morula—the developing embryo may die. Most often, this happens when the irregularity in development occurs in the first week or two after conception. In fact, about half of human embryos never make it to the fetal stage, and the woman doesn't even know that she was pregnant.

So, not all changes are viable ones. Think of a set of building blocks. They are supposed to make a castle, if you follow the plan. If you put one building block out of alignment, or leave one out, then depending on where it fits in the castle you may end up with only a minor change—in a turret, say—or you may have a castle with an unstable foundation.

As with the building blocks, each step in the development of the embryo depends at least partly on what has gone before. When all goes well, the fertilized egg cell divides and grows, groups of cells differentiate into layers of tissue, and these bulge out to form organs or limb structures and further differentiate into specialized tissues. In the process, the cells lose the ability to go back and become anything else. Genetic switches control which part of the genetic blueprint a particular cell will follow, beginning with the position in the morula. Locating specific tissues in certain body regions followed on the evolutionary innovation of segmentation: advanced worms, insects, and all vertebrates have segments. In humans, segmentation occurs when the embryo is about a centimeter long and five weeks old. After this stage, the cells are programmed to become a certain region of the body: those at one end will become the head, the next group will become the upper torso with the forelimbs, then the abdomen with the hind limbs. Some of the genes that control the development of segmentation are called homeobox genes (Hox genes, for short). Once these directions have begun, they cannot be undone, even if you move the cells around. The successful cloning in 1996 of Dolly the sheep, from the DNA of a single ewe udder cell after hundreds of tries, was scientifically important because it demonstrated the exception that proves this rule. When this experiment has been repeated, it will show that under some special circumstances a cell from an adult organism—a cell that has already differentiated—can direct the development of an embryo into a healthy organism.

Many experiments have been done to see how much potential an embryonic cell has at different stages of development. Since all of an embryo's cells carry its entire genome, understanding the switches that determine whether a cell will become an eye cell, say, or a lymphocyte is a fascinating problem. In humans, after about the fifth week, any body part designed for a segment other than the head—arms, legs, heart, liver, genitalia—will not appear on the head. Simple as that—for humans and other mammals, that is. Mutations in Hox genes or other developmental-control genes can cause enough bizarre malformations in fruit flies to delight the most avid sci-fi B-movie fan. (We'll discuss Hox genes in more detail in chapter 8.)

Similar rules apply to each body segment. A choreographed series of changes must occur, all in the right order, for normal development of the brain and sense organs in the human embryo. By the five-week mark, two dark dots have appeared in the head segment which indicate where the eyes will develop, and two small indentations on either side of the head will become the ears.

All vertebrate embryos show the same developmental stages of the face; it's likely that *Star Trek* humanoids would, too.

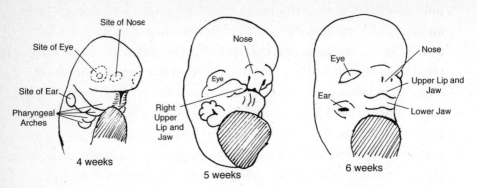

The formation of the human face. Embryos shown at four, five, and six weeks.

The next eight weeks are crucial in the development of the human head and face. If, during this time, the normal growth of brain and bone structures is impeded, birth defects usually result which affect the position or operation of the sense organs. The eyes may be spaced too closely or too far apart, or the ears may be placed too low. If normal development of the sides of the upper and lower jaws is thwarted, so that they fail to merge at the midline, a cleft lip and palate result.

Or the brain itself may be damaged. Remember that the brain and the facial bones are developing together. Problems that block or hinder the growth of one often affect the other as well. The fact that certain facial defects occur in conjunction with generalized deficiencies of intellect or motor development may explain our attraction to beautiful faces. If faces that are in some way abnormal indicate a greater chance of decreased intelligence or capability, then it is natural for us to decide, consciously or even subconsciously, that a beautiful face guarantees the opposite. Recent studies of the psychology of beauty and its relationship to the selection of mates have shown that humans from very different societies all have a strong preference for facial symmetry. If Professor Galen is right about our common ancestry, it seems highly unlikely that a *Star Trek* humanoid species would have developed an asymmetrical face, for its members would probably have come to the same conclusions that humans have—namely, that symmetry is a marker for positive traits and therefore an inducement to whatever pleasure-laced activity leads to procreation. Andorians, for example, have two antennae equidistant from their central facial axis: no problem there! And—so far, at least—the *Star Trek* makeup artists have, consciously or even subconsciously, accommodated their own species' preference for a pleasing symmetry by locating their important alien facial variations along that central axis—the Bajoran nose ridges, the Klingon forehead ridges, the Cardassian forehead and neck webbing.

Well, if alien faces are not all *that* different from human faces, then why are these slight deviations so fascinating to us, so compelling, so difficult to ignore? Perhaps the reason lies in our early cultural evolution. Before the development of language among primitive humans—certainly before passwords were invented—the face was the primary means of communication. It was how our ancestors identified and communicated with others of their own species. It therefore took on an enormous importance. We evolved a specialized ability to read the facial expressions of other humans—which may in part explain why vision became the dominant of the five human senses. Facial expressions served as signals among the members of one's own tribe—signals for help, for company, maybe even for a little territorial space in moments of stress. Indeed, facial expressions constitute a language of their own, one that sometimes reveals our true intentions more clearly than our words or actions do.

"Man is a social animal," observed the seventeenth-century philosopher Benedict Spinoza. He wasn't referring to our gift for chat; many nonverbal creatures—the elephant, the ant—are social, too. Some rely on body language to exchange information about the world. Bees dance to tell each other where the best clover is.

Others, like wolves, will glance at one another's face, as an instantaneous visual check on whether they are being threatened or the advance is a friendly one. One of Charles Darwin's lesser known works is *The Expression of the Emotions in Man and Animals* (1872), in which he notes the similarity of many animal and human facial expressions.

We are not surprised to find that among apes, whose social organization is complex, facial communication is elaborate. Primatologists have observed the nonverbal language of chimpanzees and other apes and have even learned to communicate in it. The highest of the primates, as they have in their spoken language, have greatly elaborated the vocabulary of nonverbal communication: human parents and children are well aware of the power of "a look" in discouraging errant behavior. One reason that Vulcans, the original *Star Trek* aliens, appear so unlike us—so "alien"—is the scantiness of their repertoire of facial expressions. To Terrans, Vulcans seem to have only two—the stiff semiscowl and the quizzical raised eyebrow—though there may well be a Vulcan subtlety of countenance we're just not discriminating enough to pick up.

To an even greater degree than Spock or Sarek or Tuvok, the android Data is a master of the inexpressive poker face. When he wants to, Data can leave his face totally devoid of any expression whatsoever—something that humans cannot do, since even the slightest muscle tension betrays our emotional state. The faces of "poker-faced" human beings can be "read" by detecting the muscle tension around the eyes and mouth, the responsive focusing of the pupils, the carriage of the head on the neck, the flickering of tiny muscles in the cheek, brow, and jaw. Much more information is available when the face is animated by emotion and the muscles of eyebrows, cheeks, and mouth are active. Indeed, among humans, a blank face is suggestive of drugged stupor or, in some cases, of a serious psychopathology. Psychiatrists carefully evaluate facial expression to watch for improvement in patients being treated for depression or psychosis.

Because we humans cannot instantly process "blank" facial expressions, we are unsettled by them. We are immediately and instinctively uncomfortable around such creatures as the reptilian Gorns of the original series and the Jem'Hadar of *Deep Space Nine* because of the absence of variation among individuals and the narrow range of facial expression. Our usual reaction is that we are not dealing with friendlies. Even Data is capable of unnerving us. His lack of facial expression may be why his personhood is so often questioned: if he displays no more emotion than a machine, why should we believe him sentient? Data has spent a lot of time learning to mimic human linguistic and facial expressions. His successful mimickry, not surprisingly, has gained him greater acceptance among his colleagues, even if they still wince at his lack of spontaneity. Perhaps Spock's raised eyebrow was an important element in earning him the friendship and loyalty of the *Enterprise* crew.

Humans often make broad and doubtless unjustified inferences about the character of alien races from their facial features. Because Bajorans, Vulcans, Romulans, and the Trill are physically similar to us in many respects, the human members of the *Enterprise* crews seem to assume that these races will share many of their own fundamental values and beliefs. Conversely, the Federation's humans instinctively expect the Ferengi, Klingon, and Cardassian peoples to deviate from Federation standards of morality and behavior because they have uglier or more ominous features. (Of course, let us not forget that this is theater. The *Star Trek* writers, makeup experts, and costumers use our biases to convey a lot of information about a character through his or her appearance.)

Far and away the ugliest of these aliens are the Ferengi—large-eared, beetle-browed, big-nosed, snaggle-toothed, and stunted. Their chief activity is profiteering; their culture is organized around the Rules of Acquisition. Humans perceive the Ferengi as prone to miserliness and greed. Humans infer the warlike nature of the Klingons from their long beards and hair, their heavy skulls and forehead ridges, their slanting brows, their prominent canine teeth, their glowering expressions. The faces of Klingons are angular, with pronounced bony structure, and they are larger than most of the other humanoid races. Many humans, mistakenly believing that warlike people are uninterested in the arts, are surprised to learn of the rich poetic and musical traditions of the Klingons.

The Cardassians, a reptilelike species, have webbed necks and scaly plates outlining the eyes, nose, and neck. Their eyes are small compared with human eyes. Their complexion is gray. Quite clearly, all this suggests a people incapable of loyalty, mercy, and compassion. Without a word being spoken, a human in the presence of a Cardassian will generally infer that this is a nasty race—certainly a race capable of atrocities in their conflict with and oppression of the "kinder and gentler" Bajorans, whose coloring and features so closely resemble those of the inhabitants of Earth, the human homeworld. A recent *Voyager* episode had Chakotay sacrifice his allegiance to the Prime Directive by becoming involved in a war—he sided with the appealing humanoids against their more alien enemy.

Part of why humans rely so heavily on their intuitive judgment of facial morphology is that they have a brain region specialized for facial recognition. Human newborns show a fascination for human faces from the first day of life. They will study a target object with the proportions, markings, and symmetry of a face and prefer it to targets that are unlike faces. Infants show these responses even before they have learned to connect their parents' faces with food and comfort. By eight months of age, human infants are able to distinguish their parents' faces with certainty, and they will display "stranger anxiety" when presented with new faces. By ten months, they are able to recognize their own face in a mirror—a key indicator of the psychological development of the self. Researchers in the neurosciences have identified individuals with brain damage from stroke, tumors, and head injuries who have lost the ability to recognize faces but are otherwise normal. Developmental neuropsychiatrists have identified learning disabilities in children stemming from their difficulty in apprehending such nonverbal cues as posture, speech intonation, and facial expression. Although of normal intelligence, these children have a hard time making friends and getting along in school—most likely because they are deficient in their ability to "read" and "speak" the oldest human language.

Another element in human communication—a fascinating and much less obvious one—is the subtle asymmetry between the movements of the right and the left sides of the face. Although we expect faces to be symmetrical, they are rarely perfectly so. The left hemisphere of the human brain—the region primarily responsible for sequential logic processing and language—is most closely tied to the musculature of the right side of the face, while the right hemisphere, which handles quantitative reasoning, intuition, and visual imagery, for the most part controls the left side of the face. In most humans, the brain's right hemisphere is more directly involved in apprehending social and emotional cues. Often you can detect deceit in humans by looking for a mismatch between the left and right sides of their facial expressions. Nervous humans may attempt to mask their feelings, but the left side of their face— the side driven by emotion—may give them away. This is not a foolproof method, however; many earnest and honest humans show an extreme asymmetry. Sometimes, for reasons unrelated to deceit, the left side of their brain is so busy processing data and trying to reach logical conclusions that it is unaware of what the friendly and relaxed right half of their brain is transmitting. Try covering up the right and then the left side of the face of your local politician's picture in the newspaper. Even more interesting, try using a hand mirror in front of a large mirror. With a little maneu-vering, you'll be able to see a symmetrical version of your left, and then your right, face.

WHAT MAKES A FACE ALIEN?

Coloration offers a much wider range of scientifically valid possi-bilities for diversity among aliens than the *Star Trek* writers have yet explored. Human blood is colored red by hemoglobin, a com-plex molecule with four atoms of iron at its core; the iron turns red when it is oxygenated, just as inorganic iron rusts when it com-bines with oxygen. The rosy cheeks of healthy children are an indi-cation of the well-oxygenated blood circulating in their systems. According to the early *TOS* episode "Miri," Vulcans developed oxygen-carrying molecules with a copper core, resulting in green blood. While polished copper is a warm reddish-yellow, when it combines with oxygen (copper in its "rusted" state) it develops a brown patina and then corrodes to green. It may be that this was an attempt to make Spock fit the "little green men" stereotype for aliens, since he was introduced in the first season of the original series. Later episodes dropped this convention, probably because it was discovered that human viewers do not react to the sight of green blood with the same empathic compassion that they feel at the sight of red blood. Green blood just repels us.

Melanin, which is brown, is the primary skin pigment in humans, but several other pigments exist among Earth species. Plant species contain carotene (orange), chlorophyll (green), and anthocyanin (red, blue, or purple). Earth also has many animal species in rainbow hues: amphibians, reptiles, insects, birds, and tropical fish; they are colored by pigments called pterins (varying from red to yellow) and guanins (which can be green or blue, or iridescent). Their existence in terrestrial organisms is, of course, an ultimate consequence of DNA programming, so there is no reason to think that such coloration could not be expressed in one or another alien species.

Many terrestrial mammalian species are striped, spotted, or dotted. It is our primate ancestry that has limited the expression of pigment in earthly humanoids to the typically uniform color pattern. It would be reasonable to expect that alien humanoids who evolved under environmental conditions different from ours could develop any combination of color, spots, or pigment markings— especially if this helped them to avoid predators or compete for mates. Jadzia Dax and Neelix can keep their spots.

It should also be possible for alien species to change their colors in response to the environment, as earthly chameleons do, or in response to emotional arousal, as several species of fish do. In fish cells, pigment is normally dispersed throughout the cell but organizes in clumps in response to neural signals. This accounts for the ability of some fish species to communicate their emotional state by becoming pale or dark, or changing color. There are even a few terrestrial species that are nearly transparent. The tiny glass catfish, *Kryptopterus bicirrhis*, is one example; jellyfish are another. Transparent aliens probably wouldn't be entirely invisible, but they might be translucent except for some vital organs, as long as they were also quite thin.

Nor should we dismiss the possibility of intelligent humanoid aliens that shimmer or glow in the dark. Recently, Masaru Okabe and colleagues at Osaka University, building on work by Doug Prasher of Woods Hole and Roger Tsien of the University of California at San Diego, introduced a jellyfish gene for a green fluorescent protein into a laboratory mouse species and produced a softly fluorescent mouse. This allowed the researchers to study the development of fetal mice in the dark of the mother's womb. Once they were born and their fur grew out, the mice lost their special glow.

So far, we have not examined any features of the humanoid

aliens of *Star Trek* which fly in the face of the rules of anatomy, physiology, and embryology. Let us now turn to antennae. Nothing resembling an earthly human has antennae, although many terrestrial mammalian species have whiskers—actually, long, stiff hairs with a nerve network at their base, which function (unlike human whiskers) as sensitive proximity detectors, highly specialized organs of touch. True antennae are found in the phylum *Arthropoda,* which includes insects and such crustaceans as lobsters and crabs. These antennae are jointed appendages of the head segment of the arthropod body, which function as sense organs.

It would be difficult to construct an evolutionary pathway that would result in antennae emerging from a humanoid head, such as we encounter in Andorians. However, we cannot rule out the possibility that there was, somewhere in the galaxy, an environment in which some type of sense organ emerging from the head or face gave such an enormous survival advantage that "antennae" eventually developed in a humanoid species. Without knowing a lot more about the planetary evolution of Andoria, which has generated beings with the most pronounced antennae of all the *Star Trek* aliens, we cannot sensibly speculate on the environmental pressures that may have made this dramatic departure from the humanoid norm possible.

There is a tissue in humans that could develop into something like antennae under such survival pressures. Human embryos have several segments, called embryonic pharyngeal arches, in what will become the head and neck; they supply the developing tissue for jaws and some neck organs. In fish, however, the pharyngeal arches develop into gills. Because these structures have been adapted for very different purposes, they might evolve to provide the organ substrate for the nervous tissue in an antennalike organ. This organ would most likely not emerge at the top of the head, though, but at either side of the neck. In sentient beings—those whose activities are directed primarily by the brain and not by instinct or reflexes of the nervous system—antennae that serve as important sense organs would likely be closely integrated into the brain, and thus located on or near the head.

The Andorian antennae protrude from the top of the head. They are stiff, thick, and stand upright. If this is a sense organ, it would necessitate another orifice in the skull to provide access to the brain and would probably result in a flattening or bulging of the skull's spherical shape, as is the case with the bony structures that support our ears, eyes, and mouth. What might these protruding sense organs do? In some vertebrate species of fish, like the bullhead, barbels protruding near the mouth respond to touch and taste, alerting the animal to the presence of food in the water. One could imagine Andoria as a planet with significant atmospheric variations: perhaps with extremes of temperature or barometric pressure in relatively small microclimates. Intelligent beings who traveled outside their home ecosystems could encounter potentially life-threatening circumstances, and would need to be continuously aware of atmospheric content, temperature, or barometric pressure in order to stay safe. An Andorian might be able to tell you exactly what floor of a skyscraper you were on and whether or not the weather pattern was changing, without looking out the window. This abnormal sensitivity might explain why Andorians consider it extremely rude to touch their antennae—it's probably painful to them.

The *Star Trek* writers might be well advised to explore odd sense organs that seem strange to humans but are firmly grounded in terrestrial biology. Pit vipers have infrared detectors. Many insect species see in the ultraviolet range, and many flowers manifest in this spectrum blossom colors we do not see. Some fish detect electric fields. Migrating birds navigate using Earth's magnetic field. We know of no terrestrial species that picks up radio frequencies, but stay tuned. The *Voyager* crew may yet encounter an alien with old-style TV rabbit ears!

Humans have ears, but nothing like the Ferengi. Ferengi ears are oversize and more sensitive to sound than are the ears of most of the other humanoid species. The figure shows how form follows function in facial morphology—an example of convergent evolution.

Quark, the bartender on Deep Space 9, once demonstrated his sensitive hearing by picking a lock without using a stethoscope to amplify the sound of the tumblers.

Ferengi ears also function as erotic organs, and are extremely sensitive to touch. From this we may infer that the ears of the Fer-

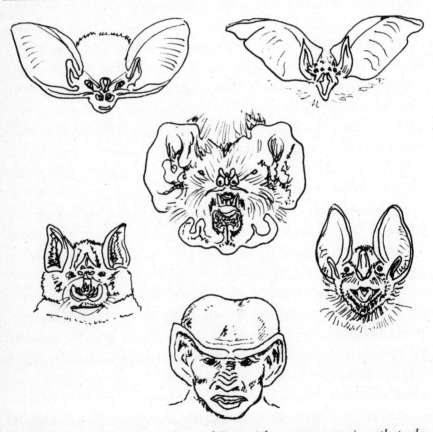

Homologies between various bat and Ferengi faces, two organisms that rely heavily on hearing.

engi are vital to their survival. The sensitivity to touch is a warning of potential peril, analogous to the extreme sensitivity of the human eye, which causes us to blink whenever a foreign object approaches; we give our eyes immediate attention when the tiniest speck of dust lands on the eyeball surface. Ferenginar, the home planet of the Ferengi, probably has an average background noise level that would seem to humans like total silence. We can guess that stiff zoning ordinances control the levels of noise in exclusive Ferengi neighborhoods. Most of their computers function with auditory rather than visual displays, unless privacy is needed. It seems likely that the Ferengi have an unusual ability to extract a particular auditory signal from a noisy background, much as humans can selectively read a single sign amid the visual confusion of Times Square. In the noisy atmosphere of his bar and casino, Quark's sensitive ears are subject to an ongoing assault, yet (as Ferengi traders are well aware) he can eavesdrop on table conversations anywhere in his establishment while appearing to be disinterestedly polishing the glasses or mopping up the countertop.

Like the Andorians, Ferengi are probably also instantly aware of changes in barometric pressure, since their eardrums and sinuses should detect and respond to these changes—though the *Star Trek* writers have yet to exploit this ability as a plot device. In conditions of high barometric pressure, Ferengi would likely experience a kind of malfunction, similar to the temporary blindness experienced by humans in the extremely bright conditions of a snowy mountaintop or sandy desert. Low barometric pressure might feel "dark" to Ferengi, since sound waves are not conducted well under such conditions. Without special training, Ferengi soldiers would likely become anxious and tentative on planets with tenuous atmospheres.

With their keen auditory awareness, one might also expect the Ferengi to have developed an interior-design esthetic that emphasizes a harmonious and beautiful auditory environment—wallpaper that sings, perhaps? For humans, interior design is an almost completely visual art. Pleasing aromas, say, or background music, while contributing a certain mood or air of refinement to a human living room or elevator or business establishment, are seldom thought of as part of the design. Humans may enjoy walking across a public space, like a city square or a hotel lobby, whose central fountain burbles and plashes; they may note as well the scent of massed flowers in the air; yet what really impresses them is what "catches the eye"—the visual excitement of a lofting space, well lit, with orderly and interesting colors and forms. But hearing is the primary sense for the Ferengi; the visual impact of such a place would be background information, and what impresses them is what "catches the ear." This may explain why Ferengi women have so complacently agreed not to wear any clothes: if visual impact is not the most important sexual signal, nakedness may not be such an important sign of female submissiveness—as it certainly would be interpreted on planet Earth. The chief sensual appeal for Ferengi is a well-presented symphony of pressure and sound waves, conveying both harmony and information. While rare (since it cannot be played for much profit), Ferengi music is truly out of this world!

Most of the sentient races in the *Star Trek* universe seem to be mammals, but reptilian races exist. On Earth, the dinosaurs might well have remained the dominant species, perhaps in time developing intelligence and even technological prowess were it not for their sudden extinction at the end of the Cretaceous, of which more later. (This extinction was thought to be the end of the story, until the *Voyager* crew encountered sentient Terran dinosaur-humanoids in the Delta Quadrant.) Among the first of the reptilian *Star Trek* races were the Gorns. In "Arena," Captain Kirk must battle a Gorn captain to the death. This Gorn is the personification of our nightmares: he, or it, is a large, scaly, gray-green, bipedal lizard with daggerlike teeth. As noted, he is expressionless (unless you consider gaping jaws a facial expression), and therefore Kirk assumes that the Gorn captain has no ability to empathize with human wants and needs. Once their representative has been vanquished by Kirk's mammalian agility and quick wit, the Gorns are seldom encountered again in the *Star Trek* story line—but they might be. Cold-blooded species are dependent on their environment to determine their metabolic rate. If the Gorns are a cold-blooded race, they would have an advantage in long-range space travel, since they could more easily adapt to the suspended animation necessary for long periods of interstellar travel by exploiting physiological mechanisms to slow down their metabolism. On the other hand, the warm-blooded races are able to maintain relatively constant high levels of physiological activity in different environments, which gives them an advantage in conflict situations. Without the ability to control the temperature of their environment—or in an environment that is unpredictable and rapidly changing, such as the one that existed on Earth at the end of the Cretaceous—cold-blooded species become vulnerable and are likely to lose out.

Cardassians are perhaps the closest thing in the canon to evolved reptiles, though they are described as "humanoid" in the *Star Trek Encyclopedia*. This disagreeable race was first encountered in *The Next Generation* and they are now a fixture on *Deep Space Nine*. They are an improbable people; on Earth, mammals have hair or fur, and reptiles and fish have scales. No terrestrial species has both, as the Cardassians appear to. This is not to say that a species with fur and scales is an impossibility. After all, egg-laying mammals should be an impossibility, but the platypus never read that textbook. Evolution seems to take delight in improbable exceptions to the rule.

It is not entirely clear whether or not the *Star Trek* writers are presenting the Cardassians as a cold-blooded race, in line with their scales. Cardassians have a prominent diamond- or tear-shaped plaque on the forehead, outlined by a scaly ridge. This might be a pineal body, an organ that is vestigial in mammals but prominently developed in some fish, amphibians, and reptiles. The pineal body is both a light receptor and a gland that produces melatonin, and it is thought essential to the establishment of circadian rhythms—those daily cycles of metabolism and hormone production we all experience. Obviously, a cold-blooded species would need to keep close tabs on the time of day; get caught in the frigid desert air after a night on the town and your metabolism might slow down to the point where you might forget to crawl home. If the Cardassians had reptilian ancestors, it might explain their preference for lower light levels and greater warmth, and their terrible taste in interior decoration (where did they find all those shades of gray for Terok Nor?).

While little has been revealed about Cardassian physiology, it is possible that Cardassians are both reptilian *and* warm-blooded. Based on the type of blood-vessel structures in fossilized dinosaur bones, and on evidence relating to heart anatomy, diet, growth rate, and various behavioral factors, some paleontologists believe that some species of dinosaurs—the fierce velociraptor, for one—may have been warm-blooded. There is evidence that edmontosaurus, a duck-billed dinosaur, tended its young.

There were no Cardassians at Camp Khitomer (unless they were spying) and Cardassia is not a part of the United Federation of Planets. Perhaps this is because a warm-blooded species that cares for its young but has a gray, scaly complexion and a cold-blooded dependence on its environment would seriously challenge humanoid diplomats of a more benign (or at least mammalian) nature. But as we shall see in the next chapter, just staying alive and well in space can present plenty of challenges no matter which world you hail from.

CHAPTER TWO

Life in Space

"Tea. Earl Grey. Hot."

—Picard, numerous times

That command, worded in the odd syntax that the *Enterprise-D*'s replicators seem to require, is as closely associated with Captain Picard as "Make it so!" Both edicts manage to convey his dignified decisiveness, and both hint at his desire (and often his ability) to instill order on a stubbornly chaotic universe.

While "Make it so!" seems to be an injunction unique to the twenty-fourth century (Captains Sisko and Janeway have uttered it, too, but Kirk never did), it expresses an ancient adherence to the military chain of command. You could easily imagine General Washington or Alexander the Great issuing the same order. But Picard's request for his favorite beverage, not to mention the *Enterprise-D*'s plush accommodations and amenities, suggest that Starfleet has recently realized that comfort is as vital to a crew's morale, performance, and good health on long spaceflights as are old-fashioned military discipline and a stout hull.

THE PERILS OF SPACE

Even by the mid-twenty-third century, when the *Enterprise* embarked on its "five-year mission," the Federation had mastered the various technologies that protect humanoids from the numerous physical dangers of extended interplanetary travel. The ship's shields and duranium hull not only deflect stray bits of space matter but also galactic cosmic rays (GCRs). Space is not a friendly place for life-forms. Supernovae are continually erupting, spewing atomic nuclei and subatomic particles—the contents of GCRs—out into the void. These particles travel with great energy and can break down molecular structures whenever they encounter them—including the proteins and nucleic acids of astronauts. Since these are subatomic particles, they aren't felt, but the damage they cause is real—cancer, infertility, and even dementia. On Earth, we are largely shielded from GCRs by our atmospheric blanket; by the time GCRs reach us, they have been slowed down by numerous collisions with atoms in the atmosphere. But the hulls of twentieth-century spaceships don't provide adequate insulation from them; so far, our shuttle astronauts and the cosmonauts on *Mir* have been relying on their relatively limited exposure time for protection.

As the twenty-first century dawns, researchers at NASA and elsewhere are wrestling with the problem of protecting astronauts and delicate equipment from GCRs as they plan for manned missions to Mars, which they hope will lift off in 2014. The round trip to the Red Planet, expected to take about three years, will expose astronauts to potentially lethal levels of radiation, and right now there isn't any material that can deflect harmful rays without bankrupting the mission. No metal will suffice, because GCRs would break up metal atoms and send the activated particles speeding through the ship, doing much the same kind of damage to human tissues as the GCRs themselves would. Perhaps some sort of newfangled ceramic composite is the answer, but with today's technology, sheathing a Mars-bound ship in such a material would eat up a major portion—at least 40 percent—of NASA's $25-billion budget for the project. Some researchers at NASA believe the solution is as simple as H_2O, theorizing that a foot-thick shell of water between the crew and the GCRs would deplete the rays of their destructive energy before they reach an astronaut's innards.

The Federation has licked another hazard that bedevils today's space explorers: weightlessness. We have all seen footage of astronauts being carried from their craft after only a couple of weeks in space. Even the most strapping space traveler quickly loses muscle tone in zero gravity, because the body's muscles, whether in the legs or in the heart, simply don't have to work hard when there's no gravity to work against. The body also loses bone mass in space; bones are living tissues that lay down more mineral at points of stress. Take away the stress and your body reabsorbs the bone mineral to use elsewhere. An astronaut may "grow" as much as an inch in space; without gravity pulling the body down, the spine lengthens.

According to the *Star Trek* Technical Manual, gravity is supplied to the *Enterprise* by a network of small gravity generators that emit gravitons. Some space stations use the older method of substituting centrifugal force for gravity by rotating the entire station. In that case, "down" is "out." On *DS9*: "Melora," we watch as Dr. Julian Bashir moves clumsily in the low-gravity environment enjoyed by Ensign Melora Pazlar, who grew up on a low-gravity planet. Although all Starfleet personnel are trained in zero-gravity conditions, the experience is rare enough that Bashir needs to be shown a few moves by his blond friend.

These environmental problems pale when compared to the psychological challenges of boredom and isolation. For submariners and people stationed in Antarctic research facilities, the absence of normal environmental cues like seasonal vegetation and changes in light, the lack of fresh and varied foods, the limited choice of recreation, and the confined quarters have been shown to cause everything from mild depression and elevated blood pressure to total breakdown. Astronauts, real and fictional, face similar challenges. The *Voyager* crew is stranded, as of this writing, 60,000 light-years from home, while Deep Space 9 is out on the contested edges of the Alpha quadrant. How do they cope?

After 400 years of experience, the Federation seems to have found solutions to most of these problems. Let's examine some of the precautions Starfleet takes to ensure that its crews don't collapse physically and mentally.

STARFLEET MARCHES ON ITS STOMACH

Astronauts of the dehydrated ice cream and Tang era must look with envy upon the sumptuous meals available to Starfleet personnel in deep space. Indeed, any harried soul of the late twentieth century would be jealous of the ease with which Deanna Troi can order up an ornate chocolate sundae or Data can feed Spot a carefully formulated bowl of Feline Supplement #52. All of this magic comes courtesy of the replicators, which can create an astonishing array of items, from uniforms to assault weapons. While it is outside the scope of this book to comment on whether such technology is feasible, we can certainly make some speculations about replicator food.

Food supplies our substance, so it is not too surprising that we are what we eat: that is, carbon, hydrogen, nitrogen, and a smattering of trace elements like iron, potassium, calcium, and zinc. The simplicity and flexibility of carbon allows it to bond in a vast variety of molecules that make up the proteins, carbohydrates, and fats that we call dinner. Most of us are familiar with the welter of artificial colors and flavors that food scientists have already synthesized. Some simpler food molecules, like sucrose (table sugar) and fructose (natural fruit sugar), can be synthesized, too, although it isn't economically sensible to mass-produce them that way instead of isolating them from food crops. Other molecules, like the ester we recognize as banana flavor, are more easily synthesized than isolated and are often used in commercial food products.

In theory, to produce a delicious meal aboard the *Enterprise* all you need are stores of organic matter, the chemical blueprint for every molecule in the favorite dishes of some hundred Federation homeworlds, and the type of energy-to-matter conversion technology that the replicators represent. Alternatively, the food producers might use nanotechnology to "build" food from its component molecules. Programming the replicators always takes place off-screen in *Star Trek*. We can guess why. Do you know, or even want to know, the component molecules of your favorite pizza? We thought not.

We imagine that programming a replicator is best done by having it scan an example of the food product you want it to produce. Then the replicator has to figure out the molecular composition of the item. With an appropriate file of the molecular composition of ingredients—like tomato sauce, pepperoni, green peppers, dough—the replicator might be instructed to produce the trend-setting pizza of the twenty-fourth century.

Of course, there are those who have nothing but disdain for the replicated bounty of Starfleet rations. Ben Sisko's father was an accomplished New Orleans chef, who deplored the food his grandson Jake was reared on aboard Deep Space 9. In "Paradise," Sisko encounters a neo-Luddite human named Alixus, who has managed to bring up her son Vinod without ever allowing him to taste replicator food. One can imagine the foodie gossip of twenty-third-century playground mothers ("I hear she made those cookies from scratch!" "Well, yes, but she replicated the frosting!"). In "Family," Picard's older brother Robert, who has stayed on Earth to tend the family vineyards, remarks contemptuously that eating replicated food weakens character. (No doubt he advises his son René that eating the breadcrusts puts hair on your chest.)

Replicator food probably doesn't taste quite like Nature's own food, because it always tastes exactly the same. We enjoy fresh foods more than packaged prepared food because some of the more volatile ingredients, including those all-important flavoring esters, break down when the food ages. Foods formed through the action of bacteria—like aged cheese or wine—would be replicated in a single stage of the aging or fermenting process, and the crew would thus miss the spectrum of flavors that occurs at different periods. The variation in flavor of fruits and vegetables that depends on soil and weather conditions would be lost. The variation in meat texture—one side of the steak grilled crisp, the other side medium rare—would be missing. There wouldn't be any crispy edge to your muffin. Depending on the programmer, or the particular food item that is scanned into the replicator, every slice of cheddar cheese would taste exactly alike—same color, same texture, same pungent odor and flavor. Replicators may produce food that is uniformly blended, "perfect." But is "perfect" what you really want, day after day, in space?

Perhaps the monotonous loveliness of a replicated diet is what prompts various Starfleet crew members to take up cooking as a hobby. Riker likes to try a new recipe from time to time; Dr. Crusher makes omelets; Sisko treats his friends to gumbo. But nobody takes a dimmer view of Starfleet replicator cuisine than the Klingons. Apparently the replicators can't do a decent *gagh*. Riker was treated to the real thing while serving aboard the *IKS Pagh* in "A Matter of Honor." He made you want to try a few stewed serpent worms yourself. On Deep Space 9, the Klingon food concession offers a welcome change from Quark's: ask for your *gagh* served writhing—nothing's more stomach turning than half-dead worms. Assuming that replicator technology on board a Klingon vessel is on a par with the Federation's, a replicated Klingon space diet might lead to mutiny among the warrior crew; we doubt very much that even the finest replicators can create life. Klingon chefs probably maintain a *gagh* hatchery and a stable of food animals, much as the sailing vessels of eighteenth-century Earth did. It's easier to keep meat fresh on the hoof. However, it would also be more labor-intensive and costly to the operation of the vessel than tapping into a supply of recycled elements. Meat on the hoof requires its own life support systems. Of course, the animal waste might provide recycled carbon compounds and water: even in the twentieth century, NASA already makes use of waste water for consumption by the astronauts.

Worf is teased by his Klingon brothers about the cushy life he leads on the *Enterprise*-D. His bed and his food are too soft. No doubt he wonders, sometimes, if it isn't Worf who has the hard life. He eats Terran foods, ready or not. His Klingon system probably took several months to accommodate the strange proteins and carbohydrates it encounters. Some people on Earth endure gas pains and cramps when they consume dairy products: they are missing a key enzyme called lactase, which is needed to digest the lactose sugar in milk products. People of African or Asian ancestry are particularly prone to lactose intolerance. Who knows what seemingly harmless Terran food would make a Klingon warrior writhe in gastric distress and cry out, "Today would be a good day to die!"

Of course, Federation nutritionists are probably hard at work on just such incompatibilities. They are also probably hard at work assembling the Minimum Daily Requirements for the hundred or more humanoid races of the Federation. With each humanoid race having evolved on a different homeworld with different animal and plant species around for food, there would be slightly different requirements for essential vitamins, fatty acids, and amino acids. The "essential" part of these minor food ingredients refers to the fact that they have to be in the diet: our bodies can't manufacture these molecules, because we lack the genetic program to produce them. All minerals are "essential," obviously, because no organism can "make" a mineral. It is likely that when Worf steps up to the replicator and orders "Lasagna with spicy sausage—Klingon variation," the program automatically deducts the lactose (or whateverose) and adds the appropriate essential amino acids to the recipe.

A TOAST TO SYNTHEHOL

Arguably, synthehol rivals the vast menu of replicated food in its morale-boosting effects on spacefarers. We have the Ferengi to thank for this remarkable elixir; they invented synthehol as a marketable alternative to alcohol. Synthehol tastes like the spirit it replaces and offers the same pleasant effects without any of the drawbacks—no hangovers! The sedative effects can be shrugged off whenever the situation calls for full alertness. The number of barroom brawls in Quark's establishment suggests that the Ferengi barkeeper tolerates impaired judgment and loosened inhibitions— behavioral changes produced by both alcohol and synthehol. Indeed, a little loss of judgment on the part of the customer can be good for business.

Is synthehol a brew too good to be true? Probably not. Aspiring Ferengis, take heed.

Alcohol (chemically known as ethanol, CH_3CH_2OH) is metabolized by the liver using an enzyme that forms acetaldehyde. The aldehyde suffix may sound familiar from your high school biology class. You recall all those pickled specimens in jars? They were soaking in formaldehyde. Acetaldehyde (CH_3CHO) isn't good for your liver, either, so your liver immediately oxidizes it into acetic acid (the acid in vinegar), which your kidneys can wash down the drain using plenty of water to flush the system clear. Aged alcoholic beverages, like wine and beer, also contain methanol (CH_3OH) in tiny amounts, and methanol is actually metabolized to formaldehyde (CH_2O) and then to formic acid. No wonder you get a hangover!

The action of acetaldehyde dehydrogenase, the enzyme that metabolizes the circulating aldehydes to the acid form that the kidneys can excrete, is the slow part of the process of ingesting, metabolizing, and excreting alcohol. Acetaldehyde dehydrogenase works at a fixed rate no matter how much alcohol there is in the system for it to work on. The average human male can handle about an ounce of alcohol an hour. Drink faster than that and you will experience the variety of physical and mental changes we know as drunkenness.

Alcohol acts at a particular receptor site on the nerve cells in your brain. This site wasn't specifically built to receive alcohol signals, even though we've had alcoholic beverages around since the dawn of agriculture. Brain researchers think it's there to receive signals from a neurotransmitter called gamma-amino-butyric acid (GABA, to its friends), a substance responsible for sedation or tranquilizing, something like an innate Valium. When a neuron's GABA site is stimulated, it opens a channel for chloride ions to come in, and their presence makes it less likely that the neuron will generate an electric potential and fire a synaptic signal to the next neuron. In other words, brain activity slows down. This can lead to partyers revving up, however, because a good deal of brain activity is actually inhibitory. If you take the brakes off-line, the action speeds up. Keep tossing back the alcohol and you go to sleep, perchance to dream (if your liver can handle the load). If you tossed back a bit too much, you won't just put your cerebral cortex to sleep but your brain stem as well. Then you stop breathing. Quark tries to switch his customers to synthehol before they get too close to that state. Dead customers spend no latinum.

So far, the outlook appears bright for the interstellar reveler, but before you reach for another synthale, be warned that Ferengi barkeepers have prevented researchers from perfecting the beverage. Alcohol is a diuretic—it flushes water out of your system. This is one reason that drinking alcoholic beverages makes you thirsty for more. (And you thought it was the pretzels!) Alcoholic beverages are mostly water, but the active ingredient, that alcohol molecule, acts on the brain to decrease the production of vasopressin, a hormone that signals your kidneys to conserve water from the bloodstream and not let it all out as urine. When vasopressin is not released by the pituitary gland, the body's need to recycle water is ignored, and you need to visit the bathroom more often. Which is a good reason to have another round. Which is why the Ferengi left that particular effect in when they made synthehol.

Vasopressin is a relatively simple protein and has been available as a prescription medication for about twenty years. Since it breaks down when ingested, it is dispensed as a nasal mist in cases of bed wetting and of diabetes insipidis, a condition in which the body does not make vasopressin in sufficient quantity to keep the fluid balance stable. Theoretically, a considerate Federation diplomatic host could pump vasopressin into the atmosphere of the banquet hall and allow all the delegates to remain comfortably seated during an evening of wining, dining, and speeches.

The more interesting puzzle about synthehol is figuring out how its intoxicating effects are quickly shrugged off. When the ship goes to Red Alert, partying Starfleet officers had better have their brains, brakes and all, back on line. With alcohol, you're stuck with a drunk crew, but what if they have been drinking synthehol? What follows is 200-proof speculation—we're aware of no actual research in this area—but, hey, we've all had a few by now, so why not?

When a Starfleet officer hears the siren and sees the red lights flash, he probably responds much the way we would. His adrenal glands put out a near instantaneous surge of adrenaline. This molecule, a.k.a. epinephrine, circulates in the bloodstream and sets up the "fight or flight" response we all experience when faced with stressful situations: increase in heart rate and blood pressure, shallow breathing, blood pumping to the central organs, dilated pupils, hair standing up. The Ferengi have only to create a molecule that stimulates the GABA receptor site in the same way that alcohol does but which will be instantaneously displaced from that site, or broken down in the bloodstream, or both, in the presence of adrenaline—and which doesn't taste absolutely foul. There you have it— a simple pharmacological problem. It shouldn't take more than a couple of decades to solve, and think of the money you'll make!

HOME IS WHERE THE HOLOSUITE IS

On a ship the size of the *Enterprise*-D, an officer need never lack for wide open spaces. In fact, each member of the *Enterprise*-D crew is allotted an average of 110 square meters (roughly 1,200 square feet) of living space. On these large Galaxy-class vessels, and at long-term postings like Deep Space 9, Starfleet has even adopted a family-friendly policy. This innovation postdates the twenty-third century; the Starfleet of Kirk, Bones, and Scotty was definitely a bachelor's profession. By the 2360s, though, crew members with families are assigned larger quarters than their single colleagues. Their children attend school on the ship, either via their own computers or in an actual classroom, depending on the size of the vessel. Crew members can personalize their own quarters in terms of decoration and structure: cabin walls are movable; holographic panels can be programmed to display favorite scenes; the computer supplies music and mood lighting at a spoken command. Moreover, perhaps inspired by the enormous popularity of the cooing tribbles, the *Star Trek* writers decided in later series to allow the crew to keep pets. The space cat, Spot, is the most famous example, but Picard keeps a saltwater aquarium, and the tribbles even show up again as pets for at least one of the *Enterprise*-D's children.

Still, the comfortable sameness of the ship would be mind-numbing after a few weeks or months. A spaceship intended to "go where no one has gone before" is likely to spend many days in isolation crossing vast stretches of empty space. Even a crew of 1,200 would begin to feel like a very small town after a few years in space.

Researchers stationed near the South Pole can probably relate to the starship experience. They are considerably less comfortable in their digs and more limited for company than the crews of the *Enterprise* or *Voyager* or Deep Space 9, but they are similarly isolated from society at large and similarly surrounded by a cold and empty expanse. These discomforts have been known to cause a range of symptoms, from the predictable depression and boredom to poor motivation and inability to focus one's eyes on objects at a distance. Antarctic residents are less prone to these symptoms when they have access to movies containing scenery more stimulating than the surrounding ice and snow.

Fortunately for the *Enterprise*-D's crew, the holodecks can supply an array of thoroughly convincing three-dimensional landscapes. The holodecks can provide ski slopes, oceans, caves, forests. Miles O'Brien even managed to dislocate a shoulder while white-water kayaking on a holodeck. Of course, not every crew member chooses to spend holodeck time in the great outdoors. Data prefers the Victorian drawing room of 221-B Baker Street, the quarters of Mr. Sherlock Holmes. Captain Picard frequents 1930s San Francisco for a bracing dose of *noir*. Worf and Alexander spend quality time cementing their father-son relationship in Earth's North American Wild West. And at least one *Enterprise*-D officer, Lieutenant Reginald Barclay, was in the habit of using the holodeck to boldly go somewhere he shouldn't have gone. The shy and bumbling Barclay would amuse himself by creating and regularly one-upping holodeck caricatures of the rest of the crew—a serious breach of twenty-fourth-century etiquette, to judge from the crew's reaction in "Hollow Pursuits." Barclay's lapses are a good illustration of what we all know: a toy like the holodeck can become something of an obsession. Humanoids have always struggled to fit into their societies. When they feel as though they're failing, they seek consolation in escapist entertainment. Of course, psychologists of the twenty-fourth century could examine 400 years of human technology addiction patterns—all the way back to the boob-tube addicts and Web heads of the late twentieth century. Deanna Troi probably had little trouble recognizing the symptoms of holodiction; her challenge lay in getting Barclay to attend group therapy sessions.

The *Voyager* crew, stranded in the Delta Quadrant, relies heavily on its holodeck to survive the long weeks between planetary stopovers. A new holoprogram is hailed as a huge event in the crew's life. Holonovels are played and replayed. Without access to Starfleet space stations to refresh their supply of entertainment, the crew has resorted to writing their own holonovels. Tuvok, much to his dismay, wrote a security simulation that became extremely popular with the crew as a holonovel, even though the characters were actual crew members. Officers can also turn to the holodecks for their own homeworld celebrations and rituals. While not as satisfying as the real thing, for a Klingon who is light-years from home the holographic Rite of Ascension can at least leave you suitably bruised and rejuvenated.

WHEN YOU'RE DOWN AND OUT

One of the most valuable attributes of the *Voyager*'s holodeck technology is the Emergency Medical Hologram. The holographic

Doctor found himself in sole charge of the crew's health after the demise of the entire *Voyager* medical staff. He became so much a part of the crew that Torres equipped him with a portable holoemitter so he could leave the holodeck and visit other parts of the ship without disappearing into thin air, a feat we are not competent to comment on. But his real bailiwick is the sickbay, and his presence there is crucial.

Starfleet encourages crew members to bring every complaint promptly to the attention of medical personnel. An outbreak of flu in the closed environment of a starship or a space outpost can spread very rapidly. Depending on their homeworld exposures, some crew members will be able to mount rapid immune responses, while others, who, like Geordi La Forge, were largely reared in the sterile environments of starships and space stations, have naive immune systems. Fortunately, Starfleet doctors have access to ventilation systems and replicators. Crew members who don't answer a summons to sickbay may find themselves escorted there by security, or the ship's doctor may override their replicator controls and spike their tea. For shipwide emergencies, medical staff can dispense various medical mists through the ventilators— for instance, an aerosol of hyronalin to combat radiation poisoning works nicely.

Doctors on Starfleet vessels are expected to function in several capacities. They are needed for clinical work, of course, and are expected to stay informed on the latest medical discoveries and treatments. But they are also charged with looking after the life-support systems on the ship, heading up the biological science research facilities, and acting as chief exobiologist on away teams. Concocting new vaccines or developing medical treatments for aliens of mysterious physiological makeup are all just in a day's work.

Of course, they have some nifty tools. The tricorder, for example, makes most of our physician friends drool. Imagine being able to wave a palm-size device over a patient and be able to discern everything from inflamed appendices to head lice! Unfortunately, the tricorder really is a bit too good to be true, but many other sickbay gadgets are right around the corner. Those beds with the continuous readouts are one example. Medical personnel in today's intensive-care units can keep track of a patient's heart rate and rhythm, blood pressure, pulse, temperature, and oxygenation status on continuous display monitors. Under special circumstances, other physiological measurements, like brain waves or intervascular pulse pressure, can also be monitored. True, our current technology involves a few hookups and wire leads, not just lying down on a special sensing mattress, but even the hookups have become surprisingly simple. Sensory pads are used under premature newborns, for example, to display pulse or temperature. Those sickbay beds are not too far away in the future.

Another frequently featured piece of *Star Trek* medical technology is the hypospray. This also exists today, although it isn't quite as elegant as Starfleet issue. Mass immunization programs, such as those sponsored by the World Health Organization, require health workers to rapidly immunize hundreds of people. It would be impossible in the bush to use individual sterile syringes and individual ampoules of vaccine. Instead, a type of air gun is used to shoot the vaccine directly through the skin. No needles are required, the instrument doesn't have to be sterilized between doses, and the vaccine is supplied from a relatively large chamber that needn't be filled for each new patient. The pressurized air tank and rubber hoses would make it a bit bulky for Dr. Crusher to lug around, though. That must be why Starfleet medical researchers developed the portable version.

Modern hospitals are also doing their best to erect force fields. Operating rooms may now be equipped with ultraviolet radiation hoods to act as continuous sterilization chambers for delicate equipment that cannot be heated. Hospital rooms housing patients with highly contagious airborne viruses like varicella (chicken pox) or tuberculosis can be depressurized, so that air is never released from the room to the outside, keeping the germs away from other patients. Other rooms are kept at a slightly higher air pressure relative to the outside, to protect the patient from airborne particles that might stray into the room. It turns out that the best protection from hostile germ takeover is still washing your hands. We've often wondered why Starfleet doctors don't seem to do that very much.

Of course, some cures are effected the old-fashioned way, even after all those centuries have passed. If you're feeling a bit peckish, sometimes nothing beats curling up with a favorite book and a cup of tea. Earl Grey. Hot.

THREE

How Alien Can You Get?

"There is a big difference between you and a virus, but both are alive."

—*Data to Dr. Farallon (TNG: "The Quality of Life")*

"I just love scanning for life-forms."

—*Data (Generations)*

*I*t's a routine mission. You have arrived at an unexplored solar system with a Class-M planet to be surveyed for possible colonization. Before the Federation establishes a colony on any new planet, it must screen the planet's biosphere and carefully check for sentient life-forms whose evolution would be disturbed by the presence of an alien race. Generally, this calls for a sensor sweep and launching a probe to the planet's surface to collect samples from soil, ocean, and atmosphere. If the findings are promising, Starfleet will dispatch a specialized science vessel to conduct a more complete survey of the planet.

The Enterprise-D is in synchronous orbit over the planet. Riker *calls the life science team to its station. Captain Picard orders a "search for life signs." As the principal investigator on this mission, you're up.*

What do you do now? How do you search for life signs?

THE SEARCH FOR LIFE SIGNS

Of course, you know that it's not sufficient to scan for movement. After all, waves in the ocean move. And there are life-forms that don't obviously show movement—trees, for example. On the other hand, if you observed movement on the planet that didn't seem to fit in with the laws of physics—say, a herd of cow-shaped things moving steadily uphill with occasional pauses that look like grazing—this might lead you to suspect that a life-form was present.

Scanning for water might be a useful way to search for life. Life on Earth is dependent on the presence of liquid water. Most earthly organisms are between 50 percent and 90 percent water. The search for extraterrestrial life in our solar system centers on Mars and on Jupiter's moon Europa, since both are thought to possibly harbor liquid water in their interiors. Simply locating liquid water on a planet won't reveal whether a life-form exists there, however. Nor should surface ice be overlooked. Many terrestrial spores, seeds, and microbes can survive suspended in ice for years, even decades. (At the other extreme, some microbes survive in geothermal pools where the temperature approaches the boiling point of water.)

What about oxygen? You wouldn't need to descend to the planet's surface (or even orbit the planet) to find out whether any was there. Like water, oxygen is easily detected by spectroscopic equipment at great distances. Earth's atmosphere developed as the result of eons of respiration by plants that fixed carbon dioxide from the atmosphere through photosynthesis and gave off oxygen as a result, so the presence of oxygen in a planet's atmosphere would strongly suggest that photosynthesis or some similar life process was going on. Oxygen isn't necessary for all forms of life, however; in fact, anaerobic bacteria are killed by oxygen. The absence of oxygen in a planet's atmosphere would not rule out life-forms—nor, for that matter, would its presence guarantee that oxygen-breathing life-forms were still there.

A better method of searching for life signs would be a search for metabolism. Life forms can be defined as self-sustaining metabolic units—systems that consume energy and raw materials, organize them, and use the products to grow and/or reproduce. In a search for metabolism, we would scan the planet for the presence of carbon, nitrogen, hydrogen, and sulfur, all of them elements involved in life processes on our own Class-M planet. If we found a sustainable chemical cycle—a solid form of carbon substance being converted to atmospheric carbon dioxide, which was then recycled into a carbon solid again—it would suggest that metabolic processes were occurring. (We'd have to be careful not to be fooled by fire, which oxidizes carbonaceous materials, thereby producing carbon dioxide, and moves and grows.)

In fact, carbon itself might be the key. All Terran organisms are carbon-based; the term "organic chemistry" refers to the chemistry of carbon compounds. Carbon is unique among the elements in having a relatively small atomic mass and tremendous flexibility in chemical bonding. It forms hundreds of thousands of compounds—more than any other element except hydrogen. This versatility results from its ability to add four bonds to its outermost shell of electrons to reach a stable energy state. The composition of the major proteins—and of the nucleic acids DNA and RNA, for that matter—can run into millions of carbon atoms, with related bonds to hydrogen, oxygen, nitrogen, and sulfur. The complexity of life is possible because of the bonding versatility of the carbon atom.

Examples of Carbon Structures

consecutive chain	branched chain	ring with one branch	a three-dimensional framework

A Neurotransmitter and an Old Friend

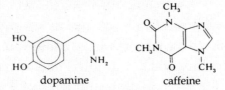

dopamine   caffeine

More Complex Organic Molecules

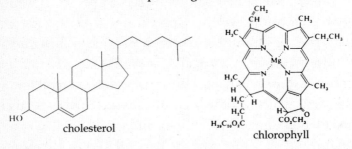

cholesterol   chlorophyll

This Is About the Best Silicon Can Do

$$-O-\underset{\underset{CH_3}{|}}{\overset{\overset{CH_3}{|}}{Si}}-O-\underset{\underset{CH_3}{|}}{\overset{\overset{CH_3}{|}}{Si}}-O-\underset{\underset{CH_3}{|}}{\overset{\overset{CH_3}{|}}{Si}}-O-\underset{\underset{CH_3}{|}}{\overset{\overset{CH_3}{|}}{Si}}-O-\underset{\underset{CH_3}{|}}{\overset{\overset{CH_3}{|}}{Si}}-O--$$

The flexibility of carbon in forming molecules.

With so many different carbon molecules to choose from, what aspect of carbon chemistry might be most useful in designing a method of detecting life signs? You'll need a method that is both sensitive (won't miss much) and specific (won't detect too many inorganic carbon compounds—for example, diamonds, or plain old graphite).

The computer requests your next direction. This is a routine scan for organic forms. You order a search for evidence of ATP hydrolysis. . . .

On Earth, virtually all living organisms use the same chemical reaction to produce energy to drive the metabolic process. In the cell, glucose is oxidized (burned) to help form adenosine triphosphate—popularly known as ATP—via a cascade of biochemical reactions. ATP stores energy in its phosphate bonds. When the energy is required by the cell, the ATP is hydrolyzed (chemically combined with water) in a stepwise fashion to trim off a phosphate group, leaving ADP (adenosine diphosphate). In some cases, the reaction goes all the way to adenosine phosphate. Among simple prokaryotic organisms (single-celled organisms without a nucleus, like bacteria and blue-green algae), the formation of ATP may take as few as ten steps. In eukaryotic (nucleated) cells, which contain mitochondria to act as the cellular power plant, more elaborate enzymatic systems are used for the production of ATP, resulting in a more efficient recovery of energy from the reaction. Some type of long-range scan that focuses on the release of energy from phosphate bonds as ATP goes to ADP would reveal the presence of actively metabolizing cells.

Another chemical signature that would be both sensitive and specific for all Earth-type life-forms would, of course, be the genetic material—the nucleic acids DNA (deoxyribonucleic acid) and RNA (ribonucleic acid). Every living creature on Earth, even viruses, contains nucleic acids as its blueprint. (Most organisms have DNA; some viruses have only RNA.) Programming the Ops console to search for the energy signature of DNA replication, DNA transcription, or RNA reverse transcription would locate organisms that are actively processing nucleic acids. Dr. Noonien Soong was aware of the Federation protocol that scanned for these reactions, and he supplied his android consort, Juliana, with a sub-processor that emitted the appropriate energy signature, in order to protect her android identity.

You order the computer to search for nucleic acids, musing as you await the results. According to the Enterprise *logs, not all life-forms are found on planets. . . .*

LIFE WHERE THE AIR IS THIN

Geordi La Forge likes to tell the story about the "space baby" that sapped the *Enterprise*-D's energy systems (*TNG:* "Galaxy's Child"). The creature's parent was killed after it attacked the *Enterprise*-D, but Dr. Crusher was able to assist in what amounted to an emergency C-section delivery of the "space baby," using the ship's phasers. The baby immediately "bonded" to the *Enterprise,* fixing itself to the hull and causing a serious energy drain. Geordi surmised that the creature fed on energy directly, appearing to pre-fer the electromagnetic frequency of 1,420 million cycles/second (mgHz) with a wavelength of 21 cm (the natural frequency of pre-cession of the spin of the hydrogen electron). Working with Starfleet designing engineer Leah Brahms to change the energy res-onance of the ship, he freed the *Enterprise*-D of the infant parasite, and the crew watched it rejoin a herd of similar organisms from a respectful distance.[1]

---

[1] Readers are referred to Lawrence M. Krauss's *The Physics of Star Trek* for a discussion of the *Star Trek* writers' confusion regarding energy frequencies in this episode.

In deep space, organisms would not feel the pull of gravity that planetary organisms feel, and there would thus be fewer structural limits on their size. On Earth, the largest land animal was the 20- to 30-ton seismosaurus, and its huge size meant a slow growth to adulthood, limited mobility, and special physiological adaptations. The modern blue whale is able to reach 150 tons because its mass is supported in water, which also facilitates its movement; but even the size of whales is limited by their ability to circulate oxygen and food molecules. We would expect a life-form that survives in deep space to be enormous—on the order of miles across—in order to filter enough of the scarce interstellar "dust" and rare photons to sustain itself. Space, while not a complete vacuum, is very nearly so. In near space—that is, the area encompassed by our solar system—hydrogen, by far the most abundant universal element, has an average concentration of only 100 atoms per cubic centimeter; in interstellar space the average concentration drops to 10 atoms/cc. In *The Physics of Star Trek*, Dr. Lawrence Krauss calculates that a spaceship traveling near light speed would need panels over 25 miles across to collect even 1 gram of hydrogen per second. But while hydrogen might be scarce in the reaches of deep space, at least it's available there. Living organisms need a variety of matter to create the sustained chemical reactions that comprise their metabolism. A deep-space organism would have to get its "stuff" from somewhere.

Light and radiant energy are also scarce in space, and some type of energy is needed to power a living organism. Starlight is available, but the low levels of radiant energy necessary to keep metabolism going would necessitate huge light-collecting leaves, like the large solar panels on satellites and space stations in Earth orbit that depend on solar power. Most plants don't thrive in the shade—they wouldn't do at all well in the perpetual night of space. Earth organisms that need a maximal surface area in order to exchange gases or collect scarce food particles generally have folds and pleats, which contain huge surface areas in small volumes (the baleen of the blue whale, with which it filters plankton from the ocean, is an example). While a pleated or folded structure might assist in filtering atoms from open space, it wouldn't help in capturing radiated energy; a structure with plates or leaves might be more feasible.

We are used to thinking of space vessels as hard—encased in thick metal hulls. But in outer space, living structures could be surprisingly delicate, since they would only rarely collide with anything. Organisms might reach the size of planets or larger. If they were low in mass because of their delicate construction, their needs for matter and energy could be kept to a minimum, but they would still have to have enough structural integrity not to rupture in the near vacuum of space. The chemical reactions that fueled its metabolism would have to be kept to a minimum as well. Since the average temperature of deep space is about 3 degrees above absolute zero (−273°C), a real space baby would have to conduct its chemical reactions in closely insulated environments and then circulate the products where needed, or else have evolved a metabolism that operates at temperatures where molecular motion all but ceases. Movement of this odd space creature through space might be entirely passive—like plankton adrift in the ocean—if the energy source was ambient radiant energy. If, however, the creature needed to gather matter with which to construct itself, or if it needed to maneuver itself toward energy sources, it would need a means of directed transportation. This would require more material and energy to maintain, and the obstacles facing our theoretical deep-space organism become insurmountable.

Big drifting filmy flakes of passive, energy-absorbing stuff, with nearly immeasurably low levels of metabolism—that's what we'd speculate might exist "somewhere out there." We would not expect to encounter ciliated herds of huge predatory footballs capable of traveling 3,000 km/sec, as Janeway and the *Voyager* crew found in the Delta Quadrant in "Elogium." And alas, appealing as they are, the enormous sentient jellyfishlike empathic organisms encountered at Farpoint Station probably won't be waiting for us when we get "out there." An organism with the evolved delicacy of structure needed to stay unfurled in the near vacuum of deep space would collapse in the atmospheric pressure and gravitational pull of a planet.

LIFE AS A ROCK

"Analysis is negative."

The computer's terse statement brings you back to your task. You shut down the assay and begin to dictate your report. But aren't you being a little premature? If Starfleet's standard operating procedure is to scan for nucleic acids and ATP hydrolysis in searching for life signs, it might explain why the crew is repeatedly fooled by life-forms not based on these compounds. . . . Like all twenty-fourth-century exobiologists, you must have heard of the Horta!

In *TOS:* "The Devil in the Dark," Captain Kirk and the *Enterprise* are called to Janus VI to investigate a series of mysterious deaths there among Federation miners. Janus VI is rich in mineral deposits. Its primitive biosphere has made it inhospitable as a place to live, but prospectors expect to put up with hardship and isolation in order to exploit valuable mineral deposits like pergium—after all, if it were easy, anybody could do it. Being melted by sulfuric acid, however, is beyond hardship, even for this tough group of men. Chief Engineer Vanderberg was worried about the rebellious attitude he sensed growing among his frightened and increasingly restive work crew.

The planet had been surveyed for life signs per Federation protocol prior to setting up the mining operation. Let's speculate

some, here: Most likely, Vanderberg had helped design the operation; he was there when they broke ground. The planet was so barren that the mining colony had almost no trouble with the algae
that often plague underground mine operations around areas of
light and moisture. And Janus VI would have had a relatively
small insect population—a real improvement over the last station
he'd worked on. . . .

What the Starfleet survey vessels had missed, and what Vanderberg hadn't thought about, was the possibility of non-carbon-
based life. Janus VI had the potential to become a Class-M planet,
but for some reason it had never developed a flourishing biosphere. A few continents on the planet supported primitive spore-
producing plants, and the oceans had enough algae growth to supply a thin oxygen atmosphere, but nothing seemed to get very big.
As a geologist, Vanderberg noted the lack of coal and oil deposits,
another indication of the scarcity of carbon-based life-forms.
What Janus VI did have was an amazing concentration of heavy
metals. Federation science teams concluded that the abundance of
heavy metals and the attendant high levels of radiation in the soil
had contributed to disrupting the evolution of life on the planet.
Not finding much else of interest, the science teams had left it at
that.

Science Officer Spock was therefore startled to find a silicon-
based creature alive in the tunnels of Janus VI (and fascinated to
find it sentient). But Janus VI is just the sort of planet on which
silicon-based life might be expected to develop—silicon being the
only other element whose atomic structure allows for the formation of the polymers, or long-chain molecules, that life requires.
We have already noted the tremendous bonding versatility of the
carbon atom, and on Janus VI carbon-based life, which would
have arisen much more rapidly and copiously than life based on
silicon, might have taken over the planet, had not the excess radiation (or whatever) discouraged its evolution.

Like carbon, silicon can also exist in a great variety of forms. One of the most common is silicon dioxide (SiO_2)—sand. The majority of minerals on Earth contain silicates. The ability of silicon to bond with oxygen and form cystal lattices that incorporate atoms of heavier elements leads to the formation of silicates in multiple manifestations of topaz, garnet, quartz (including amethyst), beryl, jade, and others more familiar to geologists and rockhounds. In combination with carbon, silicon can form rubberlike silicone polymers. But silicon forms such strong bonds with oxygen that it lacks the versatility required for metabolic reactions, where bonds must be formed and broken within the homeostatic environment of a living cell. That's why we think carbon-based life-forms would otherwise have predominated: silicon reactions occur much more slowly.

The mysterious rocklike creature on Janus VI turned out to be a Horta, the sole adult left to watch over the decades-long development of her race's eggs, a stage in the 50,000-year life cycle of the species. Since the Horta was composed mainly of silicon, she had no ATP or nucleic acids that would have revealed her presence to the scanning teams. She had the patience of stone and remained motionless (or at least unnoticed) until her nursery was threatened by the mining activities and she began her killing spree.

To support a life-form, whether it is carbon- or silicon-based, chemical processes must be carefully regulated. Terran life-forms have evolved catalysts out of proteins (enzymes), but it is just possible that a mineral-rich planet like Janus VI might have minerals available to catalyze the chemical reactions needed by a silicon-based life-form. Catalysts confer the advantage of better control over intracellular chemical reactions. This is why living organisms go to so much trouble to find and inhabit a particular environmental niche and set up sophisticated mechanisms of homeostasis to maintain oxygen, moisture, and temperature: those delicate protein catalysts need to be protected. Even Dr. McCoy's great-great-

great-great-great-great-grandmother understood the importance of controlling fevers—although she doubtless didn't think of it as "protecting the chemical reactions of cell metabolism by maintaining a homeostatic environment for ATPase and other key enzymes." No wonder the Horta became irritated when the miners brought light, humidity, and circulating air into her nursery. If her silicon-based life-form required catalysts derived from the surface of minerals to facilitate oxidation and reduction reactions, the miners could wipe out the entire Horta population by leaving oily fingerprints on the rocks. (Of course, the smashing of her eggs by the miners' drilling operations probably made her a little testy, too.)

The Horta was weird, but she reproduced and reasoned (and even flopped around) much like our carbon-based selves. There is, however, little reason to expect that a silicon-based organism would resemble Terran life-forms to this extent. On Velara III, for example, a subsurface species of life-form was discovered which consisted of a sophisticated organization of silicate crystals (*TNG*: "Home Soil"). These "microbrains," which lived in the moist soil just above their planet's saline water table, were threatened by a "terra-forming" project, which would have killed them by lowering the water table. The *Enterprise*-D crew narrowly averted planetary genocide by revealing the presence of the life-forms to the Federation terra-formers. The microbrains asked the "ugly bags of mostly water" (their way of describing humans) to please leave, and they did.

This *Star Trek* life-form makes sense, up to a point. After all, crystalline silicon chips are used to code information in computers. With a broad array of organized crystals—say, the size of a whole planet—bathed in a specialized electrolyte solution to supply contact between crystal stations, and with power supplied by sunlight, it is conceivable that a brainlike neural network capable of independent thought processes could exist. It would look like an elaborate and complex geological structure rather than a life-form, which would account for why the Federation scientists did not at first recognize the microbrains. The only problem with the scenario is that every silicon-crystal array capable of storing information and currently known to exist was put together by a sentient carbonaceous biped.

CRYSTALLINE ENTITIES

Perhaps the big question is not really how to search for life signs, but what qualifies as a life-form. While struggling to understand the microbrains, Dr. Crusher recited "the basic definition" of organic life: to be deemed alive, something must possess "the ability to assimilate, respire, reproduce, grow and develop, move, secrete, and excrete."

The Crystalline Entity, like the Horta and the microbrains, escaped detection by Starfleet searches for "new life" because it was so alien that its existence as a life-form was not clear. It was larger than the *Enterprise*-D, and its crystalline structure—a giant lattice of hexagons—suggested that it was simply an elaborate mineral formation. It absorbed energy directly from organic material, rather than ingesting and digesting the material itself, and it preyed upon the biospheres of planets, completely destroying all organic forms. This diet indicates that the Entity required carbon-based chemical reactions for its sustenance, although it did not appear to respire, secrete, or excrete.

As a crystalline structure, the Entity would have required a very stable environment in order to grow to its enormous size. Crystals form in conditions of relative chemical purity with long periods of quiet—no shaking or stirring! But there's another way for the Entity to have acquired its wonderful geometric structure. Some types of phytoplankton in the ocean do it every day.

Diatoms are microscopic free-floating algae, abundant in the oceans and cold freshwater lakes of Earth. They have chlorophyll and carotenoids, with which they photosynthesize their food. Diatoms grow their own silicate cases called frustules, which are made of hydrated amorphous silica ($SiO_2 * nH_2O$), and these cases are complex and often latticelike.

Swimming pool filters use diatomaceous earth (piles of dead diatom shells) to trap organic contaminants. Diatoms are also used in sanding products, because their hard silicate shells fragment into billions of sharp edges; silicate is basically glass. But unlike the Crystalline Entity diatoms are microscopic. The brittle structure of silicate doesn't lend itself to the growth of large organisms. In fact, diatoms get smaller with each generation, because of that rigid silicate case. When they divide, the new organism forms inside the "parent" case. After about sixty of these cycles, diatoms can't get any smaller, and they enter a sexual-reproduction phase in order to start off again at full size. (Note, by the way, that diatoms are not a silicon-based life-form: their metabolism is carbon-based; the silicate frustule is a beautiful case, nothing more.)

The Crystalline Entity exists in space, where its elaborate crystalline lattice would not be subjected to gravitational pulls or erosion due to weather. We can account for its latticelike structure either through an organic process (like the formation of a diatom shell) or a geological process (like the formation of a cave stalactite). Could the Entity be sentient? Could it communicate? In *TNG*: "Datalore," Lore calls the Entity across the enormous reaches of space, so there must be a way for it to perceive signals— to "hear." A latticelike crystalline structure would respond to being bent: since sound waves don't travel in space, Geordi uses graviton pulses in "Silicon Avatar" to produce minor deformations in the Entity and get its "attention."

There's a Terran example of crystalline distortion: Piezoelectric crystals convey tiny electrical impulses when their physical structure is distorted through applied pressure. Quartz, a silicate mineral, also displays this property. Perhaps the Crystalline Entity or something like it could use subtle physical distortions of its lattice as the basis of receiving and transmitting information. The Entity would need some sort of muscle equivalent in order to generate these distortions in structure—which would require the development of an elastic tissue, and that in turn would rely on the silicon atom's ability to form a variety of molecular structures that could stretch, bend, fold, and polymerize. And the Entity would require a silicon equivalent of DNA to encode and pass along the information to create such tissues. All this, to our way of thinking, rather eliminates the possibility of a living crystal without some auxiliary organic tissue whether silicon or carbon based. But if such a thing did exist, would it be a life-form?

The Crystalline Entity (if we are to continue to speculate about this improbable organism) is most likely alive in the sense that a virus is alive. Viruses do not move, breathe, eat, or think. Viruses cannot act voluntarily to protect their own existence. They have evolved to reproduce themselves by hijacking the systems of the host cell and getting them to follow the instructions of the viral DNA or RNA. Viruses are little more than protein envelopes for a bit of nucleic acid. With a stretch of imagination (something *Star Trek* does for you), we could hypothesize a Crystalline Entity capable of chemical reactions but with no more intention than a virus particle. The Entity encountered by the *Enterprise*-D drifts through the universe and is drawn toward planets with the energy signature that makes its structure resonate in a certain way; it absorbs all the chemical energy by breaking down every carbon bond it finds, and then moves on.

Of course, there's that gravity thing when you get near planets. . . . Really, that is such a bummer for enormous rigid crystalline life-forms!

Research by Stanley B. Prusiner, of the University of California at San Francisco, and others into the nature of prions suggests that

we might have crystalline entities much closer to home. These are
not enormous rigid silicate crystals but microscopic intracellular
proteins. Prion diseases have been known and studied for generations, always commanding respect and surrounded by mystery.
The Fore people of New Guinea practiced ritualized cannibalism,
with the brain of the dead being prepared by the women of the
tribe for consumption. Among these people, a degenerative neurological illness called kuru was observed and described by Western
physicians in the early 1900s. It was a fairly rare malady and
occurred late in life; its victims suffered from a wobbling, widespread gait, oscillating tremor, and loss of mental faculties.
Through the years, other such illnesses among humans and animals were observed and found to produce similar lesions in the
brain. Creutzfeldt-Jakob disease is one; scrapie, a disease of sheep,
is another; and in the 1980s and 1990s bovine spongiform
encephalopathy (a.k.a. "mad cow disease") sprang up in unprecedented numbers among cows in England.

After intensive research, the infectious particle was discovered
to be a protein able to induce a conformational change in the
healthy intracellular proteins of the victim. It was not a "slow
virus," as had once been thought; the prion has no genetic material at all. Rather, the infectious prion particle is a protein whose
folding has lost its way. Substitutions in the amino-acid sequence
(amino acids are what make up proteins)—which in inherited
cases of the illness are caused by mutations and in infectious cases
are caused by the prion's "teaching" the host-cell proteins the new
folding pattern—lead to storage of prions within the brain cells.
Much like a crystal directing its own growth by means of its con-
formation, prions induce other intracellular proteins to follow its
pattern and fold in the same way. After a while, neurons containing the prion deposits are simply too full to function. With enough
neurons incapacitated, the person (or sheep, or cow) becomes sick.
Prion diseases remained medical mysteries for so long because they

could be transmitted as if they were caused by infectious agents, and the infectious material (prion protein) could be destroyed by proper sterilization, but no organisms could be "cultured" from infected tissues.

The current debate is whether prions qualify as a life-form: they contain no nucleic acids; they do not utilize ATPase. Like a virus, however, a prion replicates itself using the host-cell machinery. Prions reproduce and assimilate, but they do not grow, develop, move, respire, secrete, or excrete. They are organic, but are they alive?

Riker's voice barks from your combadge. "Biolab, are you about finished in there? We've been ordered to proceed to our next mission objective." Startled into activity, you punch in orders for the computer to store its sensor recordings. Your chief technician gives you a thumbs-up—specimens from the probe have been retrieved. You'll be able to complete your analyses later.

"Yes, sir. Biolab ready to depart orbit. . . ."

THE PROBLEM OF ANDROIDS

In nearly every episode of *The Next Generation,* Data searches for life signs. It's something he loves to do, and there is a subtle irony here. The fictional *Enterprise*-D sensors can detect life-forms from a considerable distance, make accurate population counts, even detect specific signatures that reveal human crew members in a city of several million alien humanoids. But if Data were to turn the Ops console on himself, it would read negative.

Data poses a challenge to the definition of life. He (or ought we to say "it"?) is capable of independent motion and activity. He is capable of reproduction, and he once built a "daughter," although she "lived" only briefly before experiencing a failure of her positronic network (*TNG:* "The Offspring"). Data's civil rights have been tested in court—once to allow him to enroll as a Starfleet officer and later (in *TNG:* "The Measure of a Man") to determine whether or not he was "Federation property" and thus obliged to put his programming at risk by participating in the research of renowned Federation cyberneticist. Determining whether something has civil rights is a separate question from determining whether it is alive, however. Data tolerates people who (like Dr. Pulaski, the *Enterprise*-D's chief medical officer) do not experience him as alive, although he never doubts his own personhood. For Data—as it was for the holographic Moriarty of *TNG's* "Elementary, Dear Data" and "Ship in a Bottle"—the best test and the final determination is Descartes' *"Cogito ergo sum."*

CHAPTER
FOUR

Whose Brain Is It Anyway?

Or Why Parasitic Possession Is Nine-tenths of the Law

"I've seen the Captain frightened, boiling mad, sick, drunk, feverish, delirious, but never before tonight have I seen him red-faced with hysterics."

—*Scotty to McCoy, in TOS:"Turnabout Intruder"*

We're in sickbay. Dr. Beverly Crusher leans over the sedated Admiral Quinn to begin her medical examination. She must figure out what could have caused this prominent and dignified Starfleet Command officer to attack Will Riker with superhuman strength. She takes a hypospray to his neck to ensure that the sedation will last, but stops short. Leaning in to examine him more closely, she exclaims, "What is that?"

Between her fingers, a fluttering and obviously alien appendage pokes out from Quinn's skin. All of a sudden Dr. Crusher knows, *if not what's wrong with Quinn, at least why something's wrong with him.*

In the *Star Trek* universe, people don't suffer from headaches or schizophrenia. The futuristic medicine of the Federation has already found cures for these basically biochemical problems. Indeed, by the twenty-fourth century, the brain itself, a deep mystery for those of us unfortunate enough to have been born into the twentieth century, has been well studied and mapped, its secrets revealed and debated. Thus when perfectly normal members of Starfleet suddenly act out of character or seem to lose control of their mental faculties—Picard becoming irritable and taking unwarranted personal risks, in *TNG*: "The Battle"; Worf transfixed by a barber's scissors for no apparent reason, in *TNG*: "Schisms"; Deanna Troi becoming a jealous, raging harpy when the aging mother of an alien diplomat dies, in *TNG*: "A Man of the People"—Federation doctors are forced to conclude that something more sinister than a simple neurological problem is responsible. What they often find is an intense interspecies conflict.

Of the many ways to come into conflict with your neighbor, especially when he is a not-so-friendly alien species, the most frightening may be mind possession: to have your thought processes taken over by an alien being, one with no compassion for your suffering and no allegiance to who or what you are. The first Federation encounter with such extraspecies invasion, you may remember, took place on the planet Deneva, a Federation colony, where neural parasites had driven the entire population, including Captain Kirk's brother, insane (*TOS*: "Operation: Annihilate!"). Although these neural parasites looked harmless enough—like the jellyfish we sometimes see washed up on terrestrial beaches—they were capable of gaining complete control over their hosts, a process that was exquisitely painful. So much so that McCoy opted for total eradication—a wise decision but one that nearly cost Spock his eyesight. The entire experience shook the Federation.

With good cause. Anyone who has ever been infested by intestinal parasites (or helped purge such parasites from their suffering hosts) can tell you that such experiences are not for the squeamish. Parasites in the brain are unimaginably more terrifying.

But let's get real. Are we solely in the area of fiction here? Even in this century we know that getting past the human blood-brain barrier (a screening mechanism that keeps potentially harmful materials in the blood system from entering brain tissue) is never an easy matter. But this is only the beginning of the difficulties that an alien species attempting to invade your brain would encounter.

We'll start simple: one parasite, one host—could it happen? On Earth, we generally think of parasites as small, slimy creatures found in jungle streams, dunghills, and other wet, nasty places. And that description fits to some extent—many of Earth's parasites are wormlike organisms that wriggle around until they gain access to another living organism, within which they feed, grow, and reproduce. But parasites are not a particular phylogenetic classification, like bacteria or viruses or vertebrates. Parasites can be found among many forms of life; some parasites are bacteria, some are more advanced fauna, some are even plants. What defines a parasite is that it is an organism deriving its sustenance or its shelter from an unwilling host.

Why, you ask, would such organisms be "selected for?" Why should evolution produce a creature who can live only inside, or on, another creature? The answer is that parasites have a substantial competitive edge in their struggle for survival. The host animal does the heavy work, both for its own survival and that of its uninvited guest: the latter simply sits around and profits from every meal the host manages to find for itself. Can you think of a better place in which to assure yourself of an unending supply of food than the human intestine? And you never have to battle wind or rain or cold! You don't even have to provide for your waste removal. Just throw it into the river downstream and let the host carry it away from you. This may explain why there are more species of parasites on Earth than there are species of free-living organisms.

On the other hand—as Kes argued to her people in *VGR:* "Caretaker"—having all your food and shelter provided by another creature does not encourage the development of independent living skills. Parasites may have highly complex, even ingenious lifestyles, but most are rather simple and exquisitely stupid.

Take the case of the amoeba. By nature, it is not necessarily a parasite; amoebae can live independently, as shapeless, nearly transparent, single-celled organisms. While a magnificent sight when enlarged to a size that can engulf the *Enterprise,* as in the *TOS* episode "The Immunity Syndrome,"[1] this bit of protoplasm does nothing more than sit in its pond and wait to engulf any stray organic bits that float by. Occasionally, it splits in half to make two little amoebae out of one, but the highlight of its life is when it meets another amoeba and they exchange chromosomes. (That's really sexy, if you're an amoeba.) Our dumb little earthly amoeba more often is the swallowed than the swallower, since other creatures, including humans, also swim in ponds and swallow a certain amount of the water in the process. Any amoeba that makes it past the acids of the stomach (no mean feat; they do it by encasing themselves as spores) stays in the gut and engorges itself, as happy as a tribble in a grain storage bin. If enough of these foreigners set up housekeeping, they can cause their host a bad case of amoebic dysentery.

Amoebae need not be ingested to get into the body. Some forms of free-living amoebae can be swept up into your nasal mem-

---

[1] At the close of "The Immunity Syndrome," McCoy raises an existential question: Could it be that the *Enterprise* functions as part of the galaxy's "immune system," battling invading extragalactic life-forms? Was the crew's valiant struggle a part of some other destiny? Are we unwitting cogs in a bigger machine? The ever practical Kirk shrugs off the question.

branes. In rare circumstances, they may travel through your sinuses and lodge in your head. If that happens, you could end up with encephalitis, a life-threatening inflammation of the brain. Might this be a first step on the road to parasite-induced mind control? Thankfully, no. The amoeba causes such serious problems in the host not because it is a smart little parasite but because it is a dumb one. It is dumb because it disobeys the primary rule of parasitic survival: First, do no harm. After all, the host's continued good health is a guarantee of the parasite's survival. The amoeba is also dumb because it doesn't realize that the key to its success is to stay hidden. It has no means to avoid triggering the body's burglar alarm. Your body detects the invader and mounts a huge defensive immune response. In the gut, the body tries to flush out the amoebic irritation with lots of water, causing a form of diarrhea called dysentery. When the amoeba gets into the brain, a much more serious invasion, the body fights back with all-out chemical and biological warfare. Antibodies and white blood cells attack the amoeba; unfortunately, the surrounding brain cells, like innocent civilians in any war, are sometimes shoved aside or killed in the process. Like many successful armies, the human immune system hasn't managed to produce bombs smart enough to kill the enemy forces while entirely sparing its own.

So are amoebae eliminated as potential mind invaders? Let us politely say that we would not think of them as the vector of choice. But there are dozens of other strangers roaming in and on you, including certain bacteria who do not provoke rejection by your body—either because they manage to hide their alien nature from its defense forces or because they choose those parts of the body where detection is difficult. When these "smarter" organisms do no harm to their hosts, they are called commensals, and the host/invader relationship is known as commensalism—a live-and-let-live relationship. There are organisms called symbionts which actually help their hosts. Humans, for example, require certain amino acids that are made only by plants—animals can't make these amino acids themselves. Other essential amino acids are made by symbiotic bacteria in our gut.

You may have felt the effects of trying to live without such helpful bacteria. Have you ever been on antibiotics for a long time—perhaps to combat pneumonia or some other long-term illness? Even after you began to recover, you may have had diarrhea for a while, because the normal bacterial flora in your gut were also killed. Most likely, the doctor told you to eat cheese or yogurt—food products that contain live bacterial cultures—to replenish the bacterial colony in your intestine. In the meantime, your gut was vulnerable to less friendly invaders and didn't digest food as well. Similarly, women get yeast infections because of the absence of beneficial bacteria in the vagina, which normally keep the yeast organisms from multiplying.

In this regard, we have our doubts about the biofilter that is a part of the *Star Trek* transporter. What happens to all those beneficial bacteria whenever a *Star Trek* away team beams back to the ship? Take heed, *Star Trek* writers! If the transporter filters are not selectively programmed for Vulcans, Klingons, Terrans, Bajorans, and so forth, how will the biofilter know which bacteria to select in which person—either to destroy as harmful or leave in as necessary? After all, bacteria good for human beings might be embarrassingly inconvenient for Klingons—or the other way around. Let us assume that the transporter biofilters are set to remove all the bacteria from every away-team member. That would be the safest procedure as far as the ship is concerned, since the crew might encounter any number of unknown germs while exploring new life and new civilizations. If so, we can only hope that off screen (if not on screen) each returning crew member is given a special yogurt-type drink designed to restock his or her own particular brand of beneficial bacteria. Without species-specific intestinal bacteria, we can scarcely bear to imagine the scope of the diarrhea likely to follow. And you thought traveling south of the border was tricky!

Star Trek has given us an unusual case of a more complex symbiosis: the Trill. They are, of course, far more advanced life-forms than bacteria or single-celled amoebae. According to Dr. Bashir, ninety-three hours after a humanoid Trill host and its sluglike symbiont are joined, they can no longer survive apart from each other. Although each retains its own brain, central nervous system, and independent perception system, the Trill symbiont shares responsibility with the Trill host for how the joined entity behaves.[2] This is unusual in a symbiotic relationship, even in the Federation. We will return to the Trill later, but for now let's look at what can happen when an invading organism does not reside quietly in its host or share behavioral responsibility with it but actually steals control away from it.

PUPPET-MASTER PARASITES

In the *TNG* episode "Conspiracy," Captain Picard and his crew on the *Enterprise*-D uncover within a number of top Starfleet officials the presence of a powerful alien—an arthropod parasite with mind-controlling abilities. These crafty creatures crossed the void of intergalactic space and entered select (and always important) Starfleet officers through the mouth. Once inside its host, the parasite would drill into the back of the throat to enter the spinal cord,

---

[2] The *Encyclopaedia Britannica* defines the state of symbiosis broadly to encompass parasitism, commensalism, and mutualism. In parasitism, the parasite benefits from the relationship and the host suffers. In commensalism, one organism benefits and the other is unaffected. In mutualism, both organisms benefit from the relationship. The Trill relationship is an example of mutualism. Here, we will respect the traditional *Star Trek* usage and refer to the invading Trill organism as a symbiont.

tunnel up the spine, and lodge near the nape of its host's neck, puncturing a hole there so that its snorkel-like gill would have access to air. The infestation conferred total control of the hosts' behavior to the parasites, who also absorbed the hosts' memories and knowledge (although not the humanoid gift for small talk, which eventually gave the infestation away). The infestation could not be diagnosed by observance of the officers' behavior; they were still able to carry out normal Starfleet procedures, converse in technical language, and engage in martial arts—though their dietary preferences now appeared to include live meal worms.

Is such a parasitic infestation possible? Is it possible even theoretically? For a mind-controlling parasite (or a parasite of any kind, for that matter) to take over a Starfleet officer, several physiological events have to occur. First, the creature has to breach the body's exterior defenses: it must be swallowed, inhaled, or able to pierce the integrity of the skin. Most terrestrial parasites are able to get inside you because they're so small that they go unnoticed, like a single shuttlecraft approaching from behind a moon. Sometimes they squeeze through the tight phalanx of cells that make up the skin, or those lining the gut or the lungs. In the *Star Trek* universe, however, to be interesting a parasite must be large enough to be seen. Once they are big enough to be easily visible, they have to be equipped with formidable defenses, in order to avoid being eliminated by simple physical means. (Your average hamster, or even your average ant, wouldn't make much of a parasite, because when you notice it starting to feed on you, you can just pick it off and fling it away.)

After breaching the host's exterior, the puppet-master parasite[3] must somehow evade or defeat the host's immune system. Earthly parasites have evolved a variety of useful techniques for avoiding

[3] Science fiction fans will recognize this term as borrowed from Robert Heinlein's classic *The Puppet Masters*.

immune system attacks. Adult schistosomes (parasitic flatworms, or flukes, which measure from 5 to 30 millimeters long) invade the bloodstream and coat themselves with proteins from human plasma as the blood washes over them. Just as hunters will sometimes spray themselves with the scent of deer to fool the animals into accepting their proximity, so these parasites convince the human immune system that they are simply a part of the human body. Another method of parasitic concealment is to hang out where the body expects to find benign foreigners. Pinworms and tapeworms avoid exciting the body's defenses by staying in the intestinal tract, where they live with a lot of other nonhost protein material, in a neighborhood where the immune system has learned to accept alien protein forms.

A further technique, if you are a parasite, is to build an impenetrable wall around yourself. Roundworms encapsulate themselves in a thick protein coat that cannot be digested by human immune cells. It's not clear how *Puppetmasterus snorkelgillus,* as we shall call it, avoided the immune systems of the various hosts it infected, but as an arthropod it might well have opted for a thick chitin coat that would withstand attacks from the host's immune system. However, this would probably have produced a whopping allergic response in the human host: swelling, fever, shock, and collapse.

The mind-control parasite that has managed to establish the beachhead it needs to avoid being killed off by the host's immune system must next find a way to integrate itself into the nervous system of its host. This requires considerable physiological ingenuity, even for just one host species, but in "Conspiracy" *P. snorkelgillus* manages to do this to members of at least two different humanoid host species (humans and Vulcans) who have very different neurologies—an unimaginable feat.

Brain tissue is extremely delicate and highly differentiated. Only 10 percent of the human brain is made up of neurons, those specialized nerve cells that convey signals to each other through electrical and chemical means. As with any other carrier of electrical signals, nerve cells must be insulated; the insulation consists of glial cells and myelin sheaths, which take up the remaining 90 percent of brain tissue. The brain has the consistency of a thick pudding—a forgiving consistency, you might think; nevertheless, the intracranial space cannot be invaded without causing pressure on structures surrounding the invasion site. Anything that tries to take up residence inside the skull of a human being—a tumor, a blood clot, a puppet-master parasite—will compress brain tissue and may lead to coma or even death. Moreover, the central nervous system (composed of the brain and the spinal cord) is very sensitive to disturbances in its chemical or electrical environment. Invasions or infections put the nervous tissue out of balance, leading to storms of false electrochemical signaling that can manifest themselves as seizures, delirium, coma, or death.

It is therefore vital that a mind-controlling parasite—at least, one that wants a long-term relationship with its host—stay small. It must avoid adding significant volume to the cranial contents, and it must avoid disrupting the chemical and electrical balance in the brain. It must also leave intact the mechanism for transmission of specific and directed nerve signals. After all, a puppet master whose string-pulling results only in a pattern of nonsensical behavior by the host hardly makes for an interesting villain. Likewise, a puppet master whose manipulation interfered with the host's breathing would be the subject of a very short script. So this parasite must be able to put itself into the loop of all the complex brain functions, such as motor coordination and memory, while disturbing none of the functions of the brain that keep the host alive. Sounds almost impossible, doesn't it?

Surprisingly, there are examples of terrestrial parasites that seem to influence their hosts' behavior, although in relatively crude ways. Mosquitoes inhabited by the malarial parasite (genus *Plasmodium*) behave differently from noninfected mosquitoes.

They develop an insatiable appetite, feeding more often and spending more time at it, frequently engorging themselves with so much blood that they cannot fly away after they're finished. This makes sense for *Plasmodium,* because the mosquito is not its permanent host but merely a carrier—a vector, in biological terms. The true host is the animal the mosquito bites, and the longer the mosquito feeds, the more time the malaria organism has to move into its new home.

Another example of parasite-induced behavior change involves the lancet fluke, *Dicrocoelium dendriticum.* This fluke enjoys one of those truly incredible lifestyles that parasites sometimes develop. In its adult phase, the lancet fluke inhabits the mammalian bile duct—generally, in cattle, sheep, or goats—where it mates and releases its eggs, which flow downstream to the bowel. When the host mammal defecates, the eggs of the fluke are passed with the feces and at least some are ingested by the land snail, *Cionella lubrica.* Inside the snail, the ingested eggs hatch and go through an asexual reproduction cycle, greatly increasing their numbers by hundreds and thousands and developing into cercariae, the fluke's larval stage. The cercariae exit their snail host by way of the snail's slimy trail and are left on whatever plants the snail has been traversing. There in the slime the cercariae are eaten by ants of the species *Formica fusca.* Now comes the puppetmaster part.

You would expect a parasite ingested by its host to end up in the host's digestive tract—and, indeed, most of the lancet fluke cercariae do end up as cysts in the ant's abdomen. But some of them migrate through the ant's body to the subesophageal ganglion (a nerve in the ant's head), where they change the way the ant behaves. Normal ants have the sense to go home at night and get out of the cold, but these zombified ants hang out on the vegetation until the wee hours of the morning. They clamp onto the tips of grasses with their mandibles and don't move until the sun heats them up the next day. In the meantime, a fair number of them are eaten by sheep or cows or goats who are grazing for breakfast—and the parasite develops into its adult form inside the animal's digestive tract, and the cycle can begin once again.

These crude changes in insect behavior are a long way from the mind control exerted by *P. snorkelgillus*, but biologists have often seen simple behaviors evolve into very complex behaviors in a relatively short evolutionary time span. The sophisticated mind control illustrated in "Conspiracy" would require two criteria: a much greater intelligence than we find in twentieth-century terrestrial parasites, and more integration with the human nervous system than has yet been observed. We shall leave the former to the potential of evolution, but we can address the latter problem now. Let's look at three models for this type of integration to see the options available.

In the simplest case, the organism would hijack the nervous system of the host, sending impulses directly to each part of the body, like the puppeteer manipulating his puppet. A second model would involve a higher level of integration—more like the second set of controls in a driver's-ed car, wherein a few small additions to the operational design effectively cede ultimate control to the driving instructor. A third model would be even more sophisticated: more like a teenager operating a remote-control car.

Let's examine these three levels of mind control in more detail, beginning with a relatively simple example. Suppose that an extraterrestrial sentient parasite wants to communicate with Earth people and decides that the parasitic mind-control route is the way to go. This is reminiscent of the way in which the nanites—tiny robots that were set loose by Wesley Crusher in a science experiment gone wrong and eventually "evolved" into sentient beings—were able to communicate with the *Enterprise*-D crew via Lt. Commander Data, in *TNG*: "Evolution." The nanites were able to rapidly establish parasitic possession of Data because he was a voluntary host. Since both the nanites and Data were designed within the framework of Federation computer technology, they probably shared some organizational structures, and it was much easier for the nanites to manipulate Data than it would be for an extragalactic parasite to manipulate a humanoid whose physiology was both foreign and unknown.

But back to our alien parasite and its invasion of Earth: let's suppose in this case that it merely wants to make first contact with its intended host's species. What does it have to do to get a human host to wave and say, "Greetings, Earth people!"?

First, it enters the host body and places itself in the spinal cord at the nape of the neck, as we saw in the "Conspiracy" episode. It would then presumably send tendrils into the brain. The tendrils must be small enough not to disrupt the brain-volume balance, and the parasite must also avoid compressing the spinal cord with its body and thereby cause paralysis. Now all the creature has to do is integrate with billions of nerve cells.

The brain, however, works less like a telephone system than like a computer: there is no central cable for a parasite to tap into for the twin purposes of monitoring your thoughts and inserting its own signal. The brain has approximately 100 billion nerve cells, with countless integrations and cross-signals necessary for sensory perception, thought, and motor control. In order to send a sensible message from the brain to your arm to get you to wave, the parasitic puppet master first taps into the motor control circuits of the left side of the brain, which control the right shoulder, upper arm, lower arm, and hand. The forty or so facial muscles and the fourteen tongue muscles needed to speak must also be brought on line. The respiratory system needs to be brought in, too, because breath must be exhaled to produce speech. The parasite also has to access the speech center of the brain, and out of dozens of possible greeting phrases, it must determine that "Greetings, Earth people!" is the culturally correct signal for the occasion. Now it integrates the "word-find" system, sends the selected phrase to the motor speech-production circuit, then operates the cascade of delicate muscular functions necessary to produce audible and intelligible speech, while simultaneously signaling hand motion and facial expression.

As you can see, the marionette method is pretty cumbersome. In the *TOS* episode "Spock's Brain," Bones rigs a computer to run Spock's body this way, after Spock's brain is stolen. Even with the complex and sophisticated computing potential of the Federation, all Bones can manage to get Spock's body to do is walk, sit, and stay. No speech and very few other movements are generated. So you can imagine how difficult it would be for an invading parasite without the power of superadvanced computers behind it.

A mind-control parasite would probably have better luck interfacing its own nervous system directly with that of the host. Then it could simply think its own thoughts and express them by using the host's neural circuits. When you walk, you don't think about lifting each foot, putting it on the ground ahead of you, and pushing the ground back behind you before you lift the other foot, and so on. You decide to move forward and your brain translates this intention into a series of neural signals that handle the mechanics. It would be easier for a parasite if it could access these higher-level pathways and not have to operate each and every neuron and move each and every muscle itself. Our parasite could try this, but it would probably need a nervous system virtually identical to the host's, in order to carry off muscle patterning. If you were inhabited by a cockroach, and the roach decided to have you walk across the street, its own nervous system would be sending the signals required to make a six-legged creature with multiple leg joints move forward. Your legs would be utterly boggled by the marching orders they were getting. Nor would even the simple up-and-down chewing motions you make when you eat be compatible with the motions of a cockroach's multiple jointed mouth parts.

And now it's going to say, "Greetings, Earth people!"? With your lips and tongue and teeth? We think not.

But let's suppose that some advanced arthropod can "put on" the nervous system and neural circuits of the host—rather like putting on a virtual-reality glove, or like the driving instructor taking control of a car. The relationship of parasite to host would be a terrifying situation of co-consciousness for the host. The host would be helpless, since the thoughts of the parasitic mind-controller would supersede the host's own thoughts and intentions, but the host would be aware that he or she was inhabited and possessed. This co-conscious type of mind-control was experienced by Commander Chekov and Captain Terrell in *Star Trek II: The Wrath of Khan*. The Ceti eels that were placed in their ears induced a state of near-total suggestibility, making Chekov and Terrell obedient to the whims of the tyrannical and vengeful Khan. Fortunately for both Kirk and the Federation, the two unwilling hosts, conscious that their actions were being controlled, put up enough resistance to the eels to assist in Khan's defeat.

With more complete control by the parasite, the host might completely lose the ability to communicate with the outside world while still retaining consciousness. In the *TNG* episode "Power Play," Miles O'Brien, Deanna Troi, and Data are taken over by alien criminal entities who are trying to find a way off the world that has been their prison for centuries. These aliens are able to access the crew's memories—the entity that controls Miles is able to recognize his wife, Keiko, and his daughter, Molly. And after he is released, Miles tells Keiko that he could see and hear everything that happened but was unable to control any of it. In other cases, perhaps the invading organism itself would take the backseat position, aware and conscious of its shared surroundings and merely eavesdropping on the host's sensory signals. (For the most part, this seems to be how the Trill symbiont behaves in its humanoid Trill host.)

At a more devastating level of mind control, the parasite might hijack all sensory systems as well as motor control, leaving the host an isolated consciousness inside his or her own body, unable to see or hear or sense anything at all. With the loss of motor control, the infected person would be unable to signal the world outside through speech or gesture, and would be trapped inside a body that went on functioning as if its owner were fully sentient and neurologically healthy. This is truly the stuff of nightmares! Would such a sensorially deprived intelligence be aware of time passing? Would it think? What would it think?

Now let us return to *P. snorkelgillus,* as encountered by Picard and the crew of the *Enterprise*-D. This is truly a sinister bug, so vastly superior in its nervous system functioning that it can climb inside a human body and direct that body's actions like a teenager operating a video game. It can adapt to the very different physiologies of numerous humanoid hosts and integrate with nervous systems of varying organization. It makes a few trial runs to get the moves down and explore the capabilities of the new host body, much as the video-game addict must do upon shifting from Mario World to Mortal Combat. After that, *P. snorkelgillus* is off and running.

Let's be honest—this may be one of those situations where the show's writers opted for compelling drama over credible science. This parasite would have to have a nervous system so complex and an intelligence so superior that ours would seem by comparison to be on the level of a snail. Why, one must ask, would any organism so advanced bother with a parasitic lifestyle? But if we allow that there might be some reason, hidden to us, for its decision to live parasitically, we still need to deal with the ultimate problem: against such a preternatural species, it is highly unlikely that our valiant crew would have regained the upper hand by the time the episode ended. Largely by good luck, Riker and Picard manage to slay the Mother creature and without her the other parasites die. In real life, assuming that such organisms existed, every person even suspected of being a host would have to be slaughtered, because leaving any infested host alive would provide an opportunity for the parasite to breed, multiply, and reestablish its hegemony, this time with a core population that had already learned to withstand Starfleet's counterattack.

NICE GUYS

But let's not dwell on the sinister. What if the invader's intent were to observe and not to control? What if the invader were willing to share and give some benefit to the host in return for its shelter? In other words, what if our parasite were a symbiont? The Trill "slugs" (as Captain Sisko calls them) do not normally harm their humanoid Trill hosts. The symbionts pass through a series of humanoid hosts (the Trill Jadzia Dax, for example, is the eighth host for the Dax symbiont). These joined entities coexist peacefully: the host receives the benefit of the memories, knowledge, and experience that the long-lived symbiont has acquired over its various tenures; in return, the host allows the symbiont an insider's view of the new experiences that the host accrues over his or her normal life span. In *TNG:* "The Host," wherein the *Enterprise*-D crew has its first Trill encounter, Odan, the Trill ambassador, is reluctant to discuss his symbiotic state, lecturing Dr. Crusher that the Trill would no more think of discussing their dual nature than we would talk about ourselves as single beings. But the facts eventually become clear. Unjoined Trill hosts are humanoids, who live a full humanoid life. Trill symbionts are invertebrates, lacking any sort of appendages and dwelling in their homeworld's cave pools, which provide constant environmental conditions for the delicate species. According to *DS9:* "Equilibrium," approximately half the Trill population is capable of being joined to a symbiont, but joining is made available only to one in every 1,000 Trill humanoids, and only after they have been psychologically screened and rigorously trained. Given what we know of Trill symbionts, they could in theory function like a parasitic puppet master—a disembodied intelligence who uses the senses and motor control of the humanoid host with or without the host's permission. But contrary to what we find in parasitic possession, Trill symbionts do not appear to exploit their hosts.

Once joined, the nervous systems of the symbiont and the Trill humanoid integrate and the joined entity becomes a new personality. As noted, after ninety-three hours the symbiont and the humanoid cannot be separated without death to the host and severe trauma to the symbiont, who must be joined to a new host within a certain time period or it, too, will die. Is such an intertwined existence of two sentient species possible? We can speculate that the Trill humanoids and the Trill symbionts evolved adaptations that made it possible. These adaptations would have included methods of integrating their nervous systems and the development of a symbiont physiology that does not excite an immune response from the Trill host. Since only half of the Trill humanoid population is said to be capable of joining, it may well be that the symbionts do excite an immune response in some Trill—the ineligibles may simply be allergic to the symbionts. But it seems that some adaptation of the Trill humanoid nervous system must be present for a humanoid to completely accept a symbiont. Thus it is puzzling that Riker was able to host the Odan symbiont for a time after the Trill humanoid host died far from the Trill homeworld.

The *Star Trek* writers have never fully explained the integration of the symbiont and host nervous systems. Perhaps the symbiont initially supplied only an enhancement of neurotransmitters (chemicals that transmit signals among neurons) to their Trill hosts, like some version of an organic Prozac pump—either providing a source of neurotransmitter molecules or a means of balancing them in conditions where the joined entity's survival was threatened. Over the eons, this simple enhancement may have evolved into the Trill co-consciousness; the Trill symbionts would have become capable of storing memories obtained while inhabiting the host body and retaining those memories when the symbiont was separated and joined to a new Trill humanoid. In fact, if memories, as some current theories hold, are indeed stored in complex neural networks, this would be possible. The Trill humanoid's brain and the symbiont would simultaneously experience the same stimuli, and thus would simultaneously realize the same set of neural signals. The Trill symbiont would thus function something like an added disk drive on a computer: it could contain both programming and memory storage, and would be capable of transferring information to and from the Trill humanoid. Since the symbiont is spared the tasks of food gathering and survival by virtue of its parasitic lifestyle, it might have evolved over time into a more and more specialized (and less self-sufficient) form of intelligence.

Trill symbionts, though individually feeble creatures, have the potential to control an entire society. What if a Trill symbiont inhabited several hosts in a row, each time with the intention of dominating some particular planetary political force? If this plan were coordinated with other Trill symbionts in their joined state, an entire puppet-master scenario would unfold. With several lifetimes to work things out, the Trill symbiont conspirators could set in motion a series of political moves that would not come to fruition for generations, so they would appear to be natural cultural phenomena. The Trill humanoids would soon become a slave race to the superior intelligence of the symbionts. Could this be why the Trill Symbiosis Commission maintains such tight control over which symbionts are made available for joining and which Trill will be hosts?

By the way, if the idea of a symbiotic relationship of this intimacy seems totally science fictional, consider one that occurred on

Earth around 2.5 billion years ago—a symbiotic relationship that changed not just the lives of the organisms involved but the course of evolution and even the chemical composition of the planet itself.

Scientists believe that around 2.5 billion years ago a form of single-celled bacteria began to diversify into different types. One of these, the cyanobacteria, developed a fantastic evolutionary invention: chlorophyll. This molecule made it possible for the bacterium to take energy from sunlight in a much more efficient way. Through the metabolic process of photosynthesis, chlorophyll binds carbon dioxide and releases oxygen into the atmosphere. The existing life-forms (including the cyanobacteria themselves) were increasingly stressed by the accumulations of atmospheric oxygen, however. Oxygen is highly reactive—check the rust on your car—and life-forms unaccustomed to it quickly die. Of course, the change in Earth's atmosphere occurred quite slowly— over several millions of years—so there was time for the simple unicellular life-forms to develop mutations that allowed them to survive in the toxic oxygenated atmosphere. Another form of bacteria, the alpha-proteobacteria, won the prize: they evolved a metabolic process called aerobic respiration, and you care deeply about this if you breathe oxygen.

There were still plenty of unicellular organisms struggling with this life-threatening crisis, though. The atmosphere, it was a-changing. If you had once enjoyed the nitrogen-ammonia-sulfur-rich atmosphere of primordial Earth, it was now time to hop on the nitrogen-oxygen-carbon-dioxide bandwagon. Earth, it seemed, was destined for Class-M status.

So one clever unicellular creature who wasn't very good at inventing new metabolic processes did the next best thing: it phago-cytized (swallowed up) an aerobic-respiring alpha-proteobacterium and had the good sense not to digest the thing. It simply took advantage of the alpha-proteo's ability to detoxify oxygen. The rest, as they say, is history.

Every one of your cells is a descendant of this primordial event. Your cells, and those of every other organism higher on the evolutionary scale than bacteria and blue-green algae (so this includes all of the true plants and animals), are eukaryotic. Eukaryotic cells of animals have a nucleus, which packages the organism's complete DNA, and they also contain mitochondria—the descendants of those ingested but not digested alpha-proteobacteria—to carry on respiration in a compartment of the cytoplasm, where those nasty oxygen free-radicals won't be able to corrupt the cell's delicate proteins and nucleic acids.

Eukaryotic cells and mitochondria form one of the most suc-

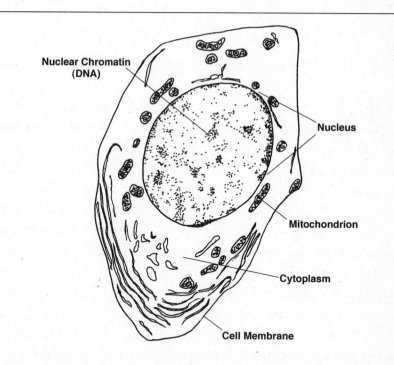

A typical eukaryotic animal cell, showing the nucleus, mitochondria floating in the cytoplasm, and cell membrane. You can see other cell structures in the cytoplasm. The nucleus is stippled; this is how chromosomes appear when they are in a resting state between reproductive cycles.

cessful commensal relationships of all time. The mitochondria have their own DNA and divide autonomously, but they have lost the ability to live independently. They are the symbionts in this story; the cell is the host. Mitochondria use the oxygen entering the host cell, giving the host energy by means of the ADP-to-ATP reaction we discussed in chapter 3, but protecting the host cell from oxygen's toxic effects. If you think this sounds amazing, you're in good company. The theory of endosymbiosis—how mitochondria and plastids (the organelles that contain chlorophyll in plant cells) were incorporated into eukaryotic cells—is one of the most exciting biological discoveries of the century. It *almost* makes *Star Trek* seem tame.

FOOLING AROUND WITH THE BRAIN

The Trill symbionts of the *Star Trek* universe are respected, even revered, but the extent of their development as independent personalities is not clear. Still, even if the symbionts are simply peripheral brains, that in itself is a fascinating concept. Could we, too, learn to link, or even network, with another entity's awareness? On Earth, we are not likely to attempt to do this organically; we are more likely to attempt it through technology. Currently, scientists can record signals from individual neurons isolated in cell culture dishes. With the recent development of nanotubules (artificially created structures only a molecule thick), scientists may soon have tools smaller than intracellular structures. The notion of tapping into individual neurons within the brain itself may soon be feasible. Moreover, several of the brain's neurotransmitters have been identified, and they are generally not complex or difficult molecules to manufacture. The most serious obstacle to technological interfacing with the human central nervous system is that the system is so huge!

The visor worn by Geordi La Forge presents us with one

instance of the interfacing of technology with the central nervous system. The visor is a truly fascinating construct, and the *Star Trek* writers have put an excellent spin on the idea by making the visor sensitive to areas of the electromagnetic spectrum other than visible light. In the *TNG* episode "The Enemy," we learn what it is like to see with Geordi's visor. The images are strange, vibrantly colorful, and distorted—a glimpse of what we might see if we could perceive infrared radiation, gamma radiation, X rays, and other energy wavelengths and particle streams outside the normally visible spectrum. The visor—perhaps not unsurprisingly—causes Geordi nearly constant headaches and occasional bouts of dizziness (both of which he considers a small price to pay for his visual abilities).

The oddness of Geordi's visual perception may also be due to his having been born blind. Kittens blindfolded at birth will grow up functionally blind even though their eyes are in working order. Without the stimulation of neural impulses from the retina, the brain's visual cortex does not develop the ability to translate light signals into meaningful interpretations of the environment. When the blindfold is removed, the signals reach the retina but the receiver in the brain is permanently off-line, and the kittens do not "see" anything. Geordi was given his visor at about five years of age and then presumably had to undergo special therapy to learn to interpret the signals that his visor was sending to his cortex. With constant practice, this interpretation skill would become automatic, much as reading is for most of us, but it would still be a learned skill rather than a hard-wired sensory system.

We are never really told how Geordi's visor interfaces with his brain, but this interface apparently makes Geordi especially vulnerable to attempts at technological mind control. In *TNG*: "The Mind's Eye," Romulan agents exploit the interface, using the visor ports as a way into Geordi's subconscious mind in order to make Geordi do their bidding. This is less an example of a possession than of a simple conditioning program—in effect, a brainwashing; the Romulans obtain Geordi's subconscious cooperation by subjecting him to increasing levels of painful stimuli. In the seventh *Star Trek* movie, *Generations*, the Klingon sisters Lursa and B'Etor, with some help from an egomaniacal El-Aurian scientist, Dr. Tolian Soran, are able to interface Geordi's visor with their ship's visual sensors, so that they can see whatever he sees. They don't exert any mind control—the aim is simply to eavesdrop. (In this movie, it seems that Geordi's vision is typical of that of other humans, because the Klingons receive a camera-perfect readout of the *Enterprise*-D's shield frequencies. However, perhaps this is because the uplink to the Klingon ship's visual sensors processed the signal impinging on Geordi's visor into its customary visual range output.)

Whether or not such eavesdropping is feasible, Geordi's visor is not so far-fetched. Serious research is currently under way to develop retinal implants based on silicon-chip technology which would convert the electrical signals from the rod-and-cone receptor cells of the eye into a signal that could be interpreted by the visual cortex of the brain. Other researchers are tackling the problem of developing artificial photoreceptors to replace deficient rods and cones. This would mean that people who have become blind through diseases of the retina, like retinal detachment or macular degeneration, could have some sight restored.

While crude by twenty-fourth-century standards, prosthetics of many kinds have long been used by humans for the accomplishment of specialized tasks. Our robotic assembly lines are just a step away from the exoskeletons and prosthetics of the Borg. Medical researchers are now able to short-circuit the central nervous system and take advantage of spinal reflexes to help paraplegics learn to walk. The spinal cord has an innate firing capacity that can stimulate muscle action by itself, without marching orders from the brain. For some paraplegics, the spinal cord can be

trained to produce a pattern of neuron firings that leads to a halt-
ing walk. This research is currently being conducted by Anton
Wernig of the University of Bonn and several other research
groups around the world. Other physiologists are developing artificial limbs that take advantage of signals from muscles that are
still intact—driving a lower-arm prosthesis with electrical impulses
from shoulder muscles, for example.

The ability to create technology that interfaces directly with the
nervous system is moving from the realm of science fiction to science fact. The next century promises to be full of miracles. We
may witness a time when the lame walk, the blind see, and the deaf
hear. "Mind control" can refer to a takeover or it can refer to the
act of using one's mind well. With the mind control of scientific
discipline and compassionate work, we can accomplish quite a lot.

In the *Star Trek* universe, of course, not all aliens are interested
in hostile takeovers. Quite a few aliens are interested in mergers of
an entirely different sort. . . .

CHAPTER
FIVE

Falling in Love with All the Wrong Faces

"I've always wanted to make love to an alien."
—*Malcorian nurse to Riker (TNG: "First Contact")*

L ove among the stars!
    The romantic customs of the humanoids of the United Federation of Planets are as varied as their peoples. Romantic complications ensnare the *Star Trek* characters nearly every week. Such a pace would exhaust mere mortals, but Kirk, Nurse Chapel, Picard, Riker, Troi, Jadzia Dax, Tom Paris, and Harry Kim, among others, have demonstrated that their romantic energies are as great as their appetites for any other type of exploration. In this chapter, we'll survey the mating rituals of some of the more familiar humanoid races, including Terrans. We'll also take a look at gender confusion, and at what happens when aliens mate: what types of humanoid hybrids might result? Finally, we'll explore a few of the more compelling encounters. Fasten your seat belt—we're about to tour the galaxy, starting with our own homeworld.

MATE SELECTION:
A GALACTIC TRAVELOGUE

Earth: Biology is a much greater determinant of our romantic inclinations here on Earth than we might at first realize. Planetwide, human males tend to choose female partners who are younger than they are, and physically healthy. Physical attractiveness is also very important to the courting male. While some standards of beauty vary from culture to culture, there is universal agreement on the beauty of a well-proportioned, healthy body. Women the world over tend to place a higher priority on financial well-being and maturity in a potential mate. The advantages with regard to the production and rearing of offspring are obvious. Men seek women in good physical condition, who will presumably be best able to bear healthy offspring. Women seek men of high status, who are able to provide food and shelter for the duration of the offspring's lengthy period of helpless dependency.

Persons of both sexes tend to favor symmetrical facial features and well-formed bodies, both of which indicate a healthy genome. Both sexes prefer persons with healthy personalities, who display an agreeable disposition, kindness, intelligence, and adaptability. Persons with physical, mental, or behavioral deformities tend to be marginalized in dating circumstances. Again, the process seems to favor the production of healthy offspring, who will themselves be able to have children and pass along their genetic inheritance. So pronounced are these courtship predilections that the purpose of sex begins to look less like ecstasy and more like getting your genes passed on via your children and your children's children. Some biologists, most notably the British neo-Darwinist Richard Dawkins, see this perpetuation of your genes as the highest imperative. In the 1989 edition of his *The Selfish Gene*, Dawkins writes that "the fundamental unit of [natural] selection . . . is not the species, nor the group, nor even, strictly speaking, the individual. It is the gene, the unit of heredity."

After these basics are noted, we Terrans tend to choose mates who are like us—not just the same religion, ethnicity, and socioeconomic class, but also similar coloring, body build, and facial features. We may be romantically attracted to an exotic alien, but most of us marry the girl or boy next door.

Vulcan: Vulcans make good scientists because they try to base decisions solely on logic. Emotions are suppressed as much as possible. Vulcan society is orderly and contemplative. Raising children, by contrast, is messy, time-consuming, expensive, and cuts seriously into your meditation time. Being logical, most Vulcans have concluded that while children must indeed be produced for the perpetuation of the race, parenthood is not a logical lifestyle choice for any individual Vulcan. To counteract this dilemma and continue to pass along their genes, the Vulcans have had to evolve an extremely powerful mating drive, the *Pon farr*.

The *Pon farr* grips Vulcan adults every seven years, creating tremendous physiological upheaval. In *TOS:* "Amok Time," Dr. McCoy notes that Spock shows an increase in "adrenal hormones" and an accelerated metabolism as the *Pon farr* overtakes him. Breathing and heart rate are up and reasoning is affected. During the *Pon farr*, Vulcans must mate or die trying. In the grip of such a powerful biological drive, the normally upright and honest Vulcans may become manipulative and devious. All their energies focus on finding a mate; every other loyalty is forsaken. Spock is irresistibly drawn homeward to T'Pring, the Vulcan female with whom he had mind-melded as a child. This telepathic bond ensures that Vulcan couples will have synchronous *Pon farr* cycles.

Vulcans are embarrassed by the *Pon farr*. Even though the drive cannot be resisted without risking death, it is considered an intensely private affair. The fastidious and dignified Vulcans would not think of discussing this matter of biology in public. On the Vulcan homeworld, vacations are probably timed to coincide with the *Pon farr*, so that work is unaffected. This tradition is more difficult to accommodate for Vulcans serving in Starfleet, especially since they are reluctant to explain their reason for requesting shore leave to their (generally human) superior officers.

There are no true Terran equivalents of the *Pon farr,* and its invention by the *Star Trek* writers seems to have been in aid of explaining how it is that Vulcans did not die out as a people—like the Essenes, the Shakers, and other human cultural groups whose members preferred celibacy to the genetic imperative. Terrans have circadian rhythms that time our daily temperature and hormonal fluctuations along with our periods of wakefulness and sleep, and female humans experience the menstrual cycle, which governs their ability to conceive. But the closest Terran examples of *Pon farr* behavior are found elsewhere in the animal kingdom. Several insect species, notably the cicada, have internally timed metamorphoses which erupt every seven, thirteen, or seventeen years. The entire generation of hibernating cicada larvae emerge in a single season to develop into the adult phase, sing their buzzing song in the trees, mate, lay eggs, and die. The biological cues that coordinate this activity are not known, though the cicada larva's diet is thought to play a role.

Considering the prominence of Vulcans as an alien race, the *Star Trek* series seldom visits the Vulcan homeworld. Little information has emerged about what it is actually like to be a Vulcan, let alone a Vulcan on a date. All we know is what we can glean from the occasional comment by Spock or Tuvok, and both of those characters, like all Vulcans, seldom discuss any aspect of their personal lives. The rampant emotions of other races, the energy spent in dealing with the emotions through discussion and argument, the need for establishing and nurturing relationships— all these seem like a waste of time to Vulcans. Humans on Vulcan are quickly frustrated in their attempts to establish intimacy. To be content on the Vulcan homeworld requires a human of extraordinary poise and self-reliance. Perhaps these were the character traits that Spock's father, Sarek, recognized in his human wife, Amanda, leading to his highly unusual selection of a human female for his mate. Spock, their offspring, is the first of several humanoid hybrids we encounter in *Star Trek*. Although in "Amok Time" he is portrayed as experiencing a full-fledged *Pon farr,* the fact that he is a hybrid makes it unlikely that the experience led anywhere. More about this problem later.

Qo'noS: Klingons, by contrast with Vulcans, are a passionate species and ready for mating at a moment's notice. Sexual talk and ribald humor are plentiful, and both sexes participate with gusto. Klingon mating rituals involve tremendous physical exertion. It is not unusual for mating couples to inflict bites and bruises; even broken bones have resulted from a night's affair. (During the Dominion siege that coincided with the engagement of Jadzia Dax and Worf, Dr. Bashir learned to expect an early morning visit from Jadzia whenever Worf was at Deep Space 9.) Foreplay typically involves an attack by the female upon the male. On the other hand, it is acceptable, and almost expected, that Klingon males will approach females with verbal or nonverbal signals expressing their sexual interest. Klingon males who force a sexual encounter, however, are scorned as lacking in honor, since Klingon males, like Terran males, are some 20 to 30 percent larger than their females.

Klingons choose their mates for life. Great respect is accorded to the marriage bond. It is the view of Klingon culture that marriage makes the Klingon couple more influential than they were as individuals. Nevertheless, divorce is not unknown, and when it happens it is quick and easy, accomplished by a simple declaration and accompanied by spitting. The Klingon wedding ceremony celebrates an ancient myth that the power of the Klingon heart, when doubled through mating, beats loudly enough to frighten the gods. Klingon wedding garments are red, the color of blood and life. Happily wedded Klingons enjoy a serenity that bachelor male and female Klingons, still in the throes of combative mate selection, long for. Probably to prevent general carnage, and also to consolidate the power of an ancestral Klingon house, Klingon mates are generally chosen for young people of high standing by their parents. The Klingon female enters the house of her intended and brings the oral tradition of her own ancestral house with her. As a new daughter-in-law, she also memorizes the ancestral lineage of her mother-in-law, and in turn, she has her own daughters and daughters-in-law memorize this lineage as well. The family's acts of honor, its titles, and its service to the society are all preserved for the edification of each succeeding generation. In this way, Klingon civilization survives despite its unbridled and often bloody mores.

Ferenginar: Among the Ferengi, sexual selection is naturally a matter of commerce. Since all of Ferengi culture is organized around the Rules of Acquisition, it is likely that marriage is viewed as another opportunity for profit. Females among the Ferengi are harshly oppressed, by Federation standards. They are not permitted to engage in business or enter into legal contracts (Rule #94: "Females and finances don't mix"), they are not to be seen in public, and they are not allowed to wear clothes. Although Ferengi marriages tend to be contractual affairs, largely arranged for the purposes of advancing the profit of both houses, the Ferengi are a promiscuous people. Mating for pleasure is exploited as a source of profit. Lip service is given to proprieties, but many (dare we say most?) Ferengi businessmen who travel keep a list of pleasure parlors, where they pursue a variety of services at a reasonable price. (Rule #223: "Beware the man who doesn't make time for *oo-mox.*")

The Trill Homeworld: The Trill apparently have sex lives similar to Terrans but with a unique twist. Because the gender of its sequential humanoid hosts can vary, the long-lived Trill symbiont can experience life both as a male and as a female. When the host dies, all contracts—including the marital bond and legal ties to children of the host-symbiont dyad—are dissolved, as if the relationships had been solely those of the host. Trill hosts are strictly forbidden to marry each other if their symbionts were married in their past lives. Symbionts are supposed to have a wide range of life experiences; the taboo is against "reassociation," and the penalty for violating it is exile, with eventual death of the symbiont when the host body dies. Nevertheless, the possibilities that a symbiont might marry his or her own offspring by a former Trill host remain. One imagines that everyone on the Trill homeworld makes a careful study of genealogy.

Rubicun III: The *Star Trek* universe seems surprisingly Victorian, on the whole. The Victorians separated sensuality from spirituality, but this aspect of Terran culture could hardly obtain all over the universe. We know that even on Earth sexuality is celebrated as an act of creation in several cultures. There is a natural association between the worship of a Creator God and the experience of sexual arousal leading to reproduction, and the process of mating and birthing may well be central to the spirituality of many races in the galaxy. On Rubicun III *(TNG:* "Justice"), the *Enterprise*-D crew was startled by the easy sensuality of the scantily clad Edo. These were a deeply religious and law-abiding people, who were carefully watched over by a spaceborne, paternalistic "transdimensional" guardian. It would be fascinating to learn more about the celebration of sexuality in this relatively uncomplicated culture. One wonders whether their marriages are monogamous, or indeed whether the Edo bother with marriage at all. In a culture that allows adults to have multiple sexual partners, children might be raised communally, and property inheritance would probably play a minor role. Unfortunately for us—but fortunately for Wesley Crusher, who trampled an Edo flower bed and nearly was made to pay with his life—the *Enterprise*-D crew left in a hurry and didn't hang around to investigate these questions.

Betazed: The Betazoids tend to mate late in life. This is probably because it is risky to mate before the full development of one's adult personality and telepathic abilities. You think it's tough living with someone who won't ask for directions or take out the garbage? Imagine being married to a telepath who ruminates on the theme music from the Betazed equivalent of *Gilligan's Island*. Terrans tend to favor pairings in which intelligence and body build are similar. Betazoids and other telepaths are much more likely to focus on traits of temperament and personality than on physique when choosing a mate, since emotional and social compatibility are much more important in their society. Lwaxana Troi notwithstanding, Betazoids probably avoid falling in love with nontelepaths. Having to explain everything in words is considered tiresome by Betazoid females; they much prefer being able to share feelings instantly, through telepathy. Betazoid women are unclothed in the traditional wedding ceremony. Nothing remains hidden for long in a marriage between two telepaths!

To ensure reproduction in this species, Betazoid women experience a quadrupling of their libido in midlife. This stage is called simply "the phase." On Betazed, women in the phase are accorded great respect, and the condition is celebrated. We can speculate that (as on Vulcan with the *Pon farr*) couples save for expensive vacations to coincide with this stage of life. Betazoid Starfleet officers might seek retirement or a leave of absence during the phase. It's just too difficult to explain to superior officers unfamiliar with this aspect of Betazoid biology. A female officer's actions during the phase, like those of the Vulcan male in the midst of *Pon farr,* may be faintly embarrassing in retrospect.

Before we consider the romantic complications of uncertain gender and interalien mating, we should make one more visit.

The Ocampa Planet: Mating among the Ocampa involves three days of sustained intercourse (!), "to ensure fertilization" *(VGR:* "Elogium"). Ocampa babies develop in an external womb on the female's back. Birthing involves rupture of this womb, which then presumably shrivels and drops off like the umbilical cord of a human infant. This arrangement makes a twisted sort of biological sense, although there are no Earth species we know of that develop an external organ to support development of an embryo. When external development does take place among terrestrial species, the embryo develops within a protective eggshell. When an animal goes to the trouble of carrying an infant to term, it is usually within a protected body space like the uterus, which confers a lot of advantages for the fetus—physical protection, stable temperature and pressure, and a germ-free environment. (Among marsupials, like kangaroos and opossums, the pouch simulates some of this; even so, reproduction among marsupials is more hazardous than among mammals, where the whole process is more protected.)

Remember Kes? Although she is now a noncorporeal life-form drifting in the Delta Quadrant, she was born Ocampan. The Ocampa live only nine years, which necessitates a very rapid development: sexual maturity, the period of *elogium,* is reached between the ages of four and five. While still on *Voyager,* Kes was prematurely thrown into the *elogium.* She was very distressed; her biological clock was whizzing on, and if she didn't mate promptly, she would permanently miss her chance at bearing a child. What the *Star Trek* writers have never explained is how the Ocampa survive as a species. If each woman gives birth only once (suggested both in "Elogium" and "Before and After") and reproduction requires two parents (male and female), the Ocampa would quickly become extinct. One way such a species might survive would be if each individual began as a female, bore at least one child, and then changed sex. In this way, reproduction would remain sexual, but each individual would mother at least one child and then go on to father at least one or more. We are not accustomed to thinking of sex change as a normal part of development, but it is a useful adaptation for many species of fish and at least one species of African frog. Among the fish species that undergo sex change, a common pattern is to begin life as a female and then change to male. Researchers have demonstrated that some of these species can reverse sex multiple times under the right conditions. One trigger might be a scarcity of members of the opposite sex; another might be the disappearance of the dominant male. Take the case of a popular home aquarium fish, *Xiphophorus helleri,* the swordtail. When there aren't enough male swordtails around, a dominant mature female will volunteer for duty, developing male sex organs and even growing the long tail fin that distinguishes the male for most aquarium keepers. These sex-changing fishes are called "successive hermaphrodites," but other fish species, like the sea bass *(Serranus tortugarum),* are simultaneous hermaphrodites. Adult sea bass produce both eggs and sperm, releasing either one or the other at each mating. The Ocampa we've observed on the *Voyager* series show typical humanoid secondary sex characteristics, so it seems more likely that they undergo spontaneous sex change than that they are true hermaphrodites.

WHO'S WHO?

You are on shore leave on Risa, a planet known for its paradisal beaches and its sensual hospitality. On Risa, there is a cultural tra-

dition of sexually welcoming strangers to the community, as part of a cultic spiritual observance that has (needless to say) promoted a thriving tourist trade. Long ago, the citizens of this charming planet conquered all galactic sexually transmitted diseases, and sex here is unaccompanied by anxiety.

So, there you are, strolling down a sunny Risan street, hoping for a close sexual encounter of the alien kind. You see an interesting creature over by the bar—nice build, lovely blue hair, graceful and athletic movements. Hmmm . . . is it male or female? Unless you know for sure, your pick-up line could sadly misfire. . . .

Speaking of wobbling gender—for most Terrans accustomed to dealing solely with other Terrans, sizing up the sex of another individual is instantaneous. We are so good at this that we enjoy the occasional confusing episode as a joke. There is a wonderful complexity to the question of whether one is a boy or a girl, however. Besides one's genetic sex (XX for women, XY for men), one also has a *core gender identity* (that is, one's sense of oneself as male or female) and a *sexual phenotype* (the male or female body build and secondary sexual characteristics). In addition, there are *gender roles*—the culturally determined window dressing that is transmitted through learning. One has to learn that boys prefer to play with cars and trucks and girls prefer to play with dolls. A great deal of contemporary psychological research is devoted to determining the degree to which sex roles are biologically based. In their controversial and scholarly 1974 book *The Psychology of Sex Differences,* Eleanor Maccoby and Carol Jackson largely disproved the then commonly held idea that boys do better in math and sciences, while girls do better in language-based subjects. The discrepancy in school performance was an artifact based on cultural bias. Subtle but real pressures from teachers and parents encourage girls to avoid showing off in math and science, while boys are strongly rewarded for accomplishments in these areas. Similarly, boys with a talent for writing are encouraged to "get out and play." Maccoby and Jackson could document only a 1 percent difference in cognitive performance between boys and girls in the many studies they reviewed. Currently, neuroscientists are moving in the other direction on this issue; there is evidence that prenatal exposure to estrogen or androgens influences fetal brain organization and will predispose the child to masculine or feminine play and behavior. The next few decades of research promise to shed both heat and light on the battle of the sexes as to who is smarter, and in what way, than whom.

On Earth, for the most part, genetic sex, phenotypic sex, core gender identity, and sexual roles are all consistent within each individual. But what if there were a humanoid species in which these layers of sexuality were dealt out in different ways? One could then have as many as sixteen sexes, although there would be only two categories relevant for procreation. Wouldn't that complicate seating at your next dinner party?

In *TNG*: "The Outcast," Will Riker (always interested in exploring the frontiers of romance) falls in love with Soren, a J'naii pilot. Among the J'naii, androgyny is the norm. Fascinated by the gender roles it encounters among the *Enterprise*-D crew, Soren allows its feminine side to come out of the closet.[1] Soren experiences itself as being female (core gender identity), but according to the episode normal J'naii do not have sexual phenotypes, genetic sex differences, or sexual roles in their culture. Soren mentions that reproduction among the J'naii is accomplished when both persons expel gametes into a pod. It is not clear whether the pod is attached to one or another of the partners. If only one of the pair

---

[1] This sensitively written and acted episode was about tolerance, not biology, but it is mentioned here as an exploration of the concept that a species which reproduces sexually does not need to develop stereotypical cultural sex roles. This is one example of how core gender identity, sexual phenotype, genetic sex, and cultural gender roles can be separated.

is physiologically capable of contributing the pod structure, then there would be genetic and phenotypic sex differences in the J'naii. If both are capable of developing the pod, however, then they are a truly hermaphroditic race, whose individuals have both the male and female function.

Contrary to normal Terran experience, sexual reproduction requires only that two mature organisms combine their genetic material to produce an individual genetically different from both of them—and that's something that can be done without a lot of role-playing. Remember the amoebae? While they usually reproduce via mitosis (doubling their chromosomes and then splitting in half), they can do it the other way, too.

MÉSALLIANCES

Now that we have briefly reviewed the mating customs on the Vulcan, Klingon, Ferengi, and other homeworlds, where the traditions of the individual races remain intact, let's explore Starfleet and the Federation. Aboard starships, on unexplored worlds, or in colonial outposts, the alien races mingle more freely. It is harder for young people to meet others of their own species and customs. Starfleet regulations encourage the mingling of the species, since all sentient beings are respected as equals; and for persons serving in Starfleet, interspecies romances and marriages have become increasingly common. The official posture of tolerance and the isolation of long space voyages foster an intermingling that citizens who remain on the home planets would probably deplore. What happens when alien humanoids meet and mate?

The mechanics of sexual intercourse between humanoids of different species would probably present only minor difficulties (with the possible exception of a murderous encounter with a randy Klingon male). Most sentient humanoids are capable of great creativity, especially when it comes to fulfilling sexual desires. It is the genetics of pairing two dissimilar sets of chromosomes which is insurmountable.

In the current era of in-vitro fertilization, one could spend a lot of time injecting sperm from one animal species into the eggs of another, but one would quickly become bored. Nothing would develop. Most animal species don't even have the same number of chromosomes, let alone enough homologous genetic material to permit fertilization and cell division. Among mammals, for example, chromosome number differs widely. Hamsters have forty-four chromosomes, dogs have seventy-eight, cats have thirty-eight. Without the same number of chromosomes, the cell division process would stop after only a few cell generations. One daughter cell would have more chromosomes than another. Neither would have an orderly array or be able to continue to divide and replicate in the thousands of cell divisions necessary to produce a viable embryo. Donkeys (sixty-two chromosomes) and horses (sixty-four) share enough genetic material so that there are usually correct genetic instructions to make a viable offspring—a mule or a hinny can develop from the combined egg and sperm. Less closely related species cannot interbreed at all.

Moreover, when a viable hybrid does develop, it is almost invariably infertile. When the unequal chromosome complement of the hybrid embryo separates during cell division in order to form the hybrid's own gametes (sperm or egg cells), there are areas of breakage and mismatching. The gametes can't form, and therefore the hybrid individual will be sterile. Whenever the occasional mule or hinny does manage to have an offspring, it is considered big news. The Romans (and we do mean Romans, not Romulans) had a saying: *"Cum mula peperit"* ("When the mule foals"), which we might render as "Once in a blue moon." Poor Spock! He went through all that *Pon farr* angst for nothing!

Humanoids being what they are, though, let us suppose that a mismatched couple like Worf and Jadzia Dax decide to boldly go where no Klingons or Trill have gone before. If they did manage to produce a little Tringon or Klill, they would have to be on the lookout for birth defects. Klingons have a redundant heart, lungs, and other vital organs, as befits a warrior race. Would the offspring of a Trill-Klingon mating have one set of lungs or two? One and a half? Equally worrisome, the offspring would probably be seriously retarded. Brain development is a delicate process, with multiple feedback steps requiring specific cellular contacts at carefully timed stages. During the formation of the brain in the first and second trimester of pregnancy, nerve cells migrate from the interior of the brain to the exterior cortex, resulting in a cortical structure of organized columns of neurons. Over many months, extensive dendritic branching develops to connect the columns of cells to each other vertically and horizontally. During the first three years of life, the branches are "pruned." In other words, during brain development, connections are sent out willy-nilly to get raw material in the right places, and then unwanted connections are later selectively eliminated. Creating and then eliminating connections in the nervous system allows for maximum flexibility in the individual's adaptation, as shaped by its experiences. Now imagine that the developing brain is getting mismatched signals from its two incompatible genomes about where the cells ought to migrate or which dendritic branches should be pruned. The effects on the brain of such a scrambled genetic program would be devastating.

Jadzia and Worf need to consider something else, too. There is likely to be a big difference between a Tringon and a Klill. The offspring of a male donkey and a female horse is a mule, while the offspring of a female donkey and a male horse (a less common pairing, because the offspring is less hardy) is a hinny. Surprisingly, mules and hinnies are quite different animals. A mule has the strong body and endurance of the ass, with the

Donkey

Horse

Mule

Hinny

Phenotypes of the species donkey and horse, and of the hybrids mule and hinny.

long legs of the horse. A hinny has a horse's body on donkey's legs.

If combining chromosomes were as simple as a random draw of half the genes from the father and half from the mother (as Gregor Mendel surmised in his pea patch), then it wouldn't matter whether the male parent was a horse or a donkey—the results would be the same. The fact that mules and hinnies are different tells us that genetics isn't that simple. It turns out that the male and female chromosome complements are not treated identically by the developing embryo. In a process known as *imprinting,* parts of both genomes are selectively inactivated when the new genome is formed, so that the male genome will govern some traits and the female genome will govern others. If the male parent is a horse and the female parent is a donkey, the offspring will be an altogether different animal from the one produced by the reverse pair. To date, only a few traits in a few selected species (for example, body size in mice) are proven to be affected by this process, but imprinting is thought to be responsible for several medical conditions.

In general, the inheritance of two genes for each trait—one from the mother and one from the father—protects the offspring: if one of the inherited genes is defective, the other will still function to produce the necessary trait. You might therefore suppose that the two genes are relatively exchangeable—that is, that either the one inherited from the mother or the one inherited from the father can function. But from what we know of imprinting, it becomes clear that some genes are more equal than others. The concept of imprinting is relevant to our understanding of several genetic syndromes—for example, Turner's syndrome, an inherited disorder resulting in short stature, infertility, and learning disabilities. The victims are always female children, but they have only forty-five chromosomes instead of the normal human complement of forty-six. Instead of being specified as XX females, they are designated with a single X, indicating that they are missing one X chromosome. The interesting part is that the severity and pattern of learning disabilities changes depending on whether the girl received her functioning X chromosome from her father or from her mother. In Prader-Willi syndrome, two copies of chromosome #15 are inherited from the mother and none from the father, and the child will nearly always be obese and mildly mentally retarded, with low muscle tone and a ferocious appetite, especially for sweets. If two copies of chromosome #15 are inherited from the father and none from the mother, a different syndrome, Angelman's syndrome, results. Children with Angelman's syndrome are usually short and skinny and moderately mentally retarded, with tense muscles. The reason for the very different results of inheriting duplicate copies of chromosome #15 is an example of the phenomenon of imprinting, but scientists still aren't sure why some genes on this chromosome (and on others) are subject to this variability in expression. The current theory is that some genes are inactivated chemically during the formation of egg cells and others are inactivated during formation of sperm cells.

Spock had a human mother and a Vulcan father. Deanna Troi is the product of a human father and Betazoid mother. B'Elanna Torres had a human father and a Klingon mother. The *Star Trek* universe is peopled with hybrids who have developed into healthy adults, manifesting traits of the species of both parents. At this point, there are not enough of these hybrids around to observe the differences that imprinting might make. For example, the product of a Betazoid father and a human mother might have entirely different traits from those of the endearing Deanna. Federation medical researchers are likely very busy collecting data on hybrids and educating Starfleet personnel about which pairings are compatible and which are not. It is conceivable that the random pairing of two alien species would lead to a mutant child who was an enormous improvement over either parent—a "golden child." For some tasks, a mule is superior to either a horse or a donkey. Deviance can be for the better, in theory. In practice, however, evolutionary pressures select for the most adaptive structures and for optimal development. Any deviance from the parental program is likely to have poor results.

Let us assume that Jadzia and Worf have the go-ahead from Starfleet medical personnel, and all hormones are ready to engage. Our last bit of advice would be that Jadzia find a good obstetrician.

Remember when Bajoran officer Kira Nerys, acting as a surrogate mother, was pregnant with Miles and Keiko O'Brien's human infant? Kira suffered from the Bajoran equivalent of morning sickness, which involves sneezing in lieu of throwing up. This actually makes loose physiological sense: pregnancy requires a delicate adjustment of the woman's immune system. In order not to be rejected as a foreign body, the fetus is encapsulated in the womb, which keeps it separate from the woman's body. Though the two bloodstreams, mother's and fetus's, are closely aligned in the placenta so that the fetus can receive oxygen and nutrients and its waste products can be carried off, there is almost no actual contact between the bloodstreams. However, the woman's immune system does undergo changes. In humans, this retuning of the immune system is thought to be partly responsible for the nausea often experienced in pregnancy. In Bajoran women, pregnancy is accompanied by hay fever symptoms—their bodies exhibit a different immune response to foreign molecules. Occasionally in human pregnancy, there is a serious immunological attack on the fetus, such as happens with Rh incompatibility. Presumably, if Jadzia becomes pregnant, Dr. Bashir will be watching for signs of immunological incompatibility.

Good obstetricians are also good mechanics. Klingon craniums are large. Dr. Bashir will probably perform an ultrasound exam several times during a hybrid pregnancy, and certainly near delivery, to see whether or not a cesarean section is advisable. The Doctor on *Voyager* could tell him a good story. In "Deadlock," he had to deliver the baby of a human mother and a Ktarian father. The baby had spiky exocranial ridges protruding from its forehead, so arranged that they would have lacerated the uterus and vagina of the mother had the Doctor not been able to manage the tricky delivery by transporter beam.

WHAT MEN AND WOMEN WANT

Hmmm. . . . What's that strange music? . . . A woman is humming eerily. . . . Violins swell. . . . A hazy light fills the screen. . . .

Enter Captain Kirk, and the love scene.

Kirk was a ladies' man, but he was also the type who could love them and leave them. He had three great loves: Edith Keeler, a social worker from Earth's past (*TOS:* "The City on the Edge of Forever"); Miramanee, the Native American priestess ("The Paradise Syndrome"), and Dr. Carol Marcus (*Star Trek II: The Wrath of Khan*), by whom he had a son. But perhaps we should not omit Starfleet attorney Areel Shaw ("Court Martial") or Ruth ("Shore Leave")—or, for that matter, the warped Dr. Janice Lester ("Turnabout Intruder"), who attempts to take over not just his ship but his body. One can't begin to count all the females of every humanoid species who have found Kirk alluring, if not irresistible. (Of course, Kirk's greatest love is the *Enterprise*. Like all naval captains, Kirk refers to the ship as "she" and quickly forsakes any humanoid female who threatens that relationship.) What are we to make of such a prolific romantic life? What has Kirk got that other men lack? Do some people have a natural charisma of the aphrodisiac sort?

One of Kirk's brief romantic liaisons was with Elaan of Troyius. This arrogant princess was a passenger aboard the *Enterprise* for a few days. As a Troyian, her tears could subdue human males and make them fall passionately in love with her. Kirk is smitten against his will, and reels like a drunkard through the painful separations and joyous reunions with Elaan. What have we here? Could Kirk be responding to pheromones?

Pheromones are airborne chemical attractants that excite a specialized olfactory organ (the vomeronasal organ) in many faunal species on Earth. This organ is neuronally wired to the brain, so when an individual encounters the pheromone peculiar to his or her species, it ignites a cascade of neuronal firing and produces an instinctive behavioral response. Some pheromones are attractors: male moths locate females across great distances. Other pheromones act to inhibit procreation: rats in crowded conditions are less fertile than normal.

Some pheromones act across species. A well-studied example is the symbiotic relationship of East African ants to acacia trees. The acacia is a tender tree, and can be damaged by insects. Territorial ants make their home on the tree and attack any insects that attempt to feed on it. Would-be grazers soon learn to leave the acacia alone. When it's time for the acacia to flower and reproduce, though, it has a problem: the ants also keep pollinating bees from landing, and this could impair the tree's ability to reproduce. The acacia flower blossoms exude a pheromone that both repels ants and attracts bees. The bees then pollinate the blossoms, and after the blossoms fade and fall off, the ants return to their guard-posts.

If a pheromonal aphrodisiac were to exist among humanoids, tears would be a probable source. Tears contain tiny amounts of chemical neurotransmitters. They are shed during periods of emotional arousal and in response to autonomic stimuli. One can easily imagine that a humanoid species might secrete a neuroactive peptide hormone with sexual-attractant activity in its tears. Of course, the most likely response by a human to an alien airborne protein would be sneezing—not very romantic, but most of biology is messy that way. The other problem is that Terrans lack a vomeronasal organ. The best Kirk can do is enjoy a sensory feast—no pheromones for us.

Alas, nothing like the power of Troyian tears exists on Earth—at least, not for humans. There are no pharmacological mind-controlling aphrodisiacs, natural or synthetic, capable of overcoming the decision-making power of the human brain. We sacrificed our instinctive behaviors eons ago, when we took a branch of the evolutionary tree that explored the power of independent individual cortical processing. It is conceivable that some type of aphrodisiac might be intoxicating, and that in a stuporous haze the subject, like Kirk in thrall to Elaan, would be highly suggestible. Isn't that how alcohol works? But so far, despite centuries of scientific and shamanistic research, no true love potion has been found. The most important human sex organ is that beige puddinglike mass between our ears. If the brain does not consent, sexual arousal does not occur.

Captain Picard is more mature and generally more reserved than Kirk in matters of the heart, but he nearly succumbed to a different type of irresistible female—Kamala, the empathic metamorph of the Kriosian people (*TNG:* "The Perfect Mate"). As a metamorph, Kamala has a unique empathic ability: she can perceive what a male desires in a mate and become that person, transforming herself into the ideal of whomever she is with at the moment. At a certain phase in her biological development, she reaches full maturity and permanently acquires the traits desired by the male to whom she has psychologically bonded. Isn't this what men and women really want? To be deeply understood and joyfully accepted by an intelligent partner who finds meaning and fulfillment in life by pleasing you? Now, there's an aphrodisiac!

CHAPTER
SIX

Not in Our Stars, but in Our Genes?

"One thing is for sure; you will never again look at your hairline in the same way."

—*Picard to Jason Vigo (TNG: "Bloodlines")*

*B*ok, *the Ferengi ex-DaiMon, is bent on revenging himself against Jean-Luc Picard, whom Bok holds responsible for his son's death. He has decided to pay Picard back in kind. Bok will search the galaxy for a young man of plausible background and age, fool Picard into believing that this impostor is a son he never knew, and then kill the "son" off. On Camor V, Bok's operatives turn up a likely candidate—Jason, the son of Miranda Vigo, an old flame of Picard's. Unfortunately, Jason's DNA doesn't match Picard's, but Bok surreptitiously resequences it. . . .*

Jason Vigo is not what Picard expected a son of his would be like—but then Picard isn't yet certain that this person really is his son. The young man paces Picard's quarters, restlessly, picking up and toying with equipment he wasn't trained to handle. His eyes dart about; is he looking for a way to escape, or casing the place for something to steal? His social skills leave a lot to be desired.

Picard tells himself to be patient. This is a civilian, and Picard isn't accustomed to dealing with civilians. And undoubtedly Jason's life on war-torn Camor V has been difficult. Picard keeps his reservations to himself as he escorts Vigo to sickbay. He will soon have a definite answer. . . .

Dr. Crusher's professional manner betrays no emotion as she takes a blood sample from Vigo and feeds it into the DNA analyzer. A few moments later, avoiding Picard's eyes, she turns and addresses Vigo directly. "Your genetic code is a cross between the DNA of your mother, Miranda Vigo, and your father, Jean-Luc Picard."

A CASE OF MISTAKEN IDENTITY

Star Trek's medical officers perform DNA analysis as quickly and easily as your doctor checks your cholesterol count. In this chapter, we will take a close look at the genetics of *Star Trek*. Some perfectly respectable genetics is presented in *Star Trek*, but the field is moving ahead rapidly. It's likely that in twenty-five years we will look back at the genetics portrayed on *The Next Generation, Voyager,* and *Deep Space Nine* with the bemused tolerance with which we now view Scotty's "overloaded transistors" and blinking computer consoles.

Could Bok "resequence" somebody's DNA, in effect "forging" a genome? Would Dr. Crusher be able to spot the forgery? The episode, "Bloodlines," hinges on the techniques of genetic analysis and genome manipulation. Dr. Crusher's task was to decide if Picard and Vigo are father and son. Presumably, she had some or all of Picard's DNA sequence on file for purposes of comparison.

The most thorough level of DNA analysis she could perform would be to sequence all of Vigo's complete DNA, or genome, which consists of some 6 billion "letters"—the bases adenine, guanine, cytosine, and thymine (usually referred to as A, G, C, and T for short). These bases are linked in complementary pairs along the long double-helical molecule of DNA, and the genes themselves are nothing more than particular sequences of these bases.

Because A is always paired with T and C is always paired with G, one spiral of the double helix is the counterpart of the other. The sequence of bases on one strand is complementary to the other, so that when the cell divides, each spiral becomes a template for the formation of a new double helix. One of the better known sentences in all of biology is the one written by James Watson and Francis Crick in their 1953 announcement (in the British journal *Nature*) of the structure of the DNA molecule: "It has not escaped our notice that the specific pairing we have postulated immediately suggests a possible copying mechanism for the genetic material."

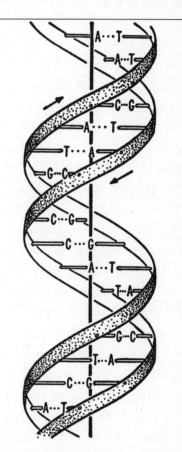

A particular gene is "expressed" when its sequence of bases is transcribed onto RNA, which then leaves the cell nucleus and enters the cytoplasm, where it translates the genetic message into a protein.

The DNA molecule. The spiral backbones are identical and consist of deoxyribose sugars connected to phosphate groups; the core consists of pairs of bases, with adenine always connecting to thymine and guanine always connecting to cytosine.

Each gene codes for one protein. This is the central dogma of molecular biology. Your complete genome is present in most of your cells. Assuming that each base pair—each "step" on the "spiral staircase" in the figure—represents a byte of computer memory, this means that one of your cells stores the equivalent of 6 gigabytes of genetic information. If Dr. Crusher had tried to analyze all of Vigo's DNA, she would have been hunched over her lab bench for an awfully long time.

The goal of the Human Genome Project, which got officially under way on October 1, 1990, is to sequence the entire human genome by the year 2005. Mapping each of 150,000 or so human genes is an enormous undertaking—not as thrilling to watch as the Moon landing or the first space shuttle launch, perhaps, but every bit as historic and important for the future of humanity. Individual genes typically vary from about 1,000 to 100,000 bases in length and some can be even longer, so sequencing even a single gene is a major accomplishment. The project is not sequencing the DNA of any one particular individual; the genome will instead be sequenced by collating the findings of many research teams in several countries. However, despite the intensity of effort over the past several years, only about 3 or 4 percent of the human genome has been sequenced at this writing. Automated systems for determining sequences in small stretches of DNA do exist, but one instrument can analyze only about 1,000 base pairs in six hours or so.[1] Even in the twenty-fourth century, analyzing someone's entire genome is likely to be a huge task.

Twentieth-century physicians can determine paternity by using

---

[1] Several researchers and commercial biotechnology firms are working on other techniques to speed up DNA sequencing. With current methods, if you were to operate 100 automatic sequencing machines continuously and in parallel, it would still take [(6 billion base pairs/100 analyzers/4,000 base pairs per analyzer per day)]/365 days per year roughly 41 years to complete the sequencing of a single human genome.

genetic analysis, however, they don't have the ability to sequence the entire genome. And Dr. Crusher has determined that Picard is Jason's father. How did she do it?

To decide whether two people are related and how closely, or to match an individual to a sample of DNA found at a crime scene, geneticists take advantage of the many polymorphisms (variants) available in the human genome. When you study human beings, it is clear that if they were products coming off an assembly line the quality control would be terrible. There is a lot of variability. In terms of phenotype—the physical result, or "expression," of the genome—we vary in height, body build, facial features, coloring, intelligence, coordination, and many other traits. This is so in great part because our genotype—the information coded in our genes—also varies from person to person. One obvious and important consequence of this genetic variability is the need for careful tissue typing and matching in organ transplantation. (Compare this situation with that of the cheetah. Cheetahs can donate organs to each other with relative ease; they show little tissue rejection. Cheetahs are nearly monomorphic— that is, they are nearly genetically identical. Scientists speculate that some cataclysm occurring between 100,000 and 200,000 years ago all but wiped out the species, with the result that modern cheetahs are very closely related to each other.)

Many of the polymorphisms used in DNA analysis are individual variations in short sequences of DNA. Only 10 percent of your genome consists of the genes themselves. The other 90 percent includes simple repeat sequences of bases, like CTCTCT (for cytosine/thymine) or CAGCAGCAG (for cytosine/adenine/guanine)— sequences that lie in between (and sometimes right in the middle of) your genes. Some of these repeat sequences are functional—for example, they may act as "switches" to turn particular genes on or off—and some are known as "junk" (an unfortunate phrase that probably means only that such sequences are not yet understood). In a typical forensic DNA analysis in the late twentieth century, geneticists examine a series of a dozen or so of these and other kinds of genetic polymorphisms in, say, a blood sample found at a crime scene. With a number of polymorphisms to choose from, and a knowledge of how often a polymorphism occurs in the general population, a geneticist can narrow down the search for suspects in a Holmesian fashion. Old Sherlock often used a number of details to build a picture of a suspect—for example, a right-handed, middle-aged, cigar-smoking man with a tweed coat and a decided limp. Only a few individuals would fit this profile, so he and Dr. Watson could be quite certain which suspect was their man. The forensic geneticist can do much better. By the time the polymorphic DNA picture is constructed, only one in a billion or one in a trillion people will fit it.

To determine paternity (or other relatedness) it is necessary only to compare a few polymorphisms. We think Dr. Crusher probably followed a Starfleet protocol for determining paternity and asked the sickbay computer to run the standard polymorphism-matching program.

And now to Bok's "resequencing" of Vigo's DNA: just how plausible is that? First of all, he would know that resequencing Vigo's entire genome was unnecessary for establishing paternity. What he needed to know was which polymorphisms were examined as part of Starfleet's paternity or identity analysis. Bok probably didn't have much trouble finding out Starfleet's protocol for determining paternity: the friendly Federation actually gives away such information! The more difficult part of his task would be to obtain the record of Picard's identifying polymorphisms. Because Picard was a Starfleet officer, we think it's likely that a record of his polymorphisms would be on file with Starfleet, in case his remains ever needed to be identified. Bok may have solved his problem by bribing some technicians at a starbase.

Next, Bok would have had to enlist a genetic technician to resequence Vigo's relevant polymorphisms. For Jason Vigo to pass as Picard's son, the resequencing would have to take imprinting into account. As we discussed in chapter 5, some genes received from the mother are inactivated, and so are some received from the father. The results of the Human Genome Project will greatly simplify research into this preferential expression of maternal and paternal genes, so by the twenty-fourth century, imprinting ought to be clearly understood; thus Bok's technician will know enough to be able to resequence *only* Vigo's paternal DNA, and will need to resequence *only* those polymorphisms that would typically be examined as part of the paternity determination.

The next problem is how to get these resequenced pieces into Jason Vigo. Twentieth-century scientists can insert a new gene or set of genes into the nucleus of embryonic cells. This is done by incorporating the new genetic material into the DNA of a simple vector (or carrier), such as a plasmid or a virus, which in turn inserts its DNA into the nucleus of the embryonic cell. The embryonic cells are rapidly dividing, and they include the new genetic material with their own genetic complement as they do. Thus the new DNA will be inherited by all the succeeding cell generations of the organism. In this way, genetic researchers have created so-called "transgenic" mice—mice that carry an altered gene, or a gene from another species, in their own genome. (Such animals are useful in the laboratory—for instance, a transgenic mouse might contain a gene for muscular dystrophy, which would allow researchers to investigate possible treatments.)

Jason Vigo was not an embryo when Bok discovered him, however. While it is possible to implant new DNA sequences into the nucleus of an adult cell, it would be impossible to introduce planted genetic polymorphisms into every cell of Vigo's full-grown body. To resequence the DNA of every one of Vigo's 10^{14} cells would require inserting a vector into such odd corners of Vigo's body as the cells of his big toe, not all of which are actively dividing. We trust to Bok's Ferengi instinct to do only what's necessary to accomplish the job and no more (Rule of Acquisition #3: Never spend more for an Acquisition than you have to). For practical reasons—in the twenty-fourth century as today—that means making use of the white blood cells.

Blood cells are constantly being produced in your bone marrow. Like embryonic cells, stem cells in the marrow can differentiate: they will become red cells or platelets or any of the various types of white blood cells. Red blood cells and platelets are not used in DNA analysis, because they have neither nuclei nor a full complement of DNA, but white blood cells do. Plenty of white blood cells are present in a small blood sample; they are the logical choice for DNA analysis, and also convenient to resequence.

Placing new polymorphisms into the DNA of Jason Vigo's white blood cells would require a series of steps, but it is theoretically possible. We think Bok's technician did something like this: A portion of Vigo's bone marrow was harvested and the stem cells were mixed, in a laboratory dish, with plasmids carrying the new DNA sequences. The cells that took up the plasmid vectors would be reinserted into Vigo's marrow. You may wonder how how the technician knew which of the cells had accepted the new DNA. This can be determined in a number of ways—for instance, by introducing into the new DNA a sequence for an enzyme resistant to a chemical that you have mixed into the cell culture medium. Therefore any cells that have not taken up the new DNA will die off.

A more serious technical difficulty is how to get this new DNA into the right locations in the genome. If it isn't in the right places, it could interrupt an important gene and cause a lethal mutation. The problem of how to put new genes into the right place on the genome is a key obstacle in the development of genetic therapies. A process called *homologous recombination* takes advantage of the known sequences of DNA that surround a particular gene. The vector DNA is implanted with bits of these known sequences, to help the new gene plunk itself down where it's supposed to be in the genome, but this is a chancy technique.

One further hurdle stands between Bok and the success of his twisted plot. Since Jason Vigo's white blood cells must carry Picard's polymorphisms (and are also carrying those of Miranda, Jason's mother), Bok has to get rid of the corresponding polymorphisms from Jason's real father. You can't have three parents! Because only a fraction of Jason's bone-marrow stem cells are likely to be resequenced, it will be necessary to kill off all of Jason's remaining original bone marrow, so that in the end all of Jason's circulating blood cells will contain the new DNA sequences. In current bone-marrow transplant procedures, radiation and chemotherapy are administered in high doses to kill the bone marrow. Until the new bone marrow grows enough red cells, white cells, and platelets to replace what was lost, the patient is vulnerable to infection and bleeding problems. At present, patients recovering from bone marrow transplants require specialized hospital wards. We're guessing that by the twenty-fourth century, bone-marrow transplantation may be a lot simpler—an outpatient procedure. The tricky part for Bok is how to give Jason Vigo a complete autologous (self-derived) bone-marrow transplant without his knowledge. For an ingenious Ferengi, perhaps this is not insurmountable: he might simply arrange for Jason to have a nasty accident that would keep him unconscious for a couple of weeks.[2]

The plot works! Picard accepts Dr. Crusher's analysis and sets about developing a relationship with the troubled and resistant young man. But almost as soon as Picard discovers paternal feelings within himself, Dr. Crusher approaches him with devastating news: Jason has a serious degenerative neurological disorder called Forrester-Trent syndrome.

We checked our medical data bases and couldn't find Forrester-

---

[2] Some of our more techy Trekker friends have pointed out that the genome resequencing might have been accomplished using nanites. We'll stick to biology.

Trent syndrome (FTS), so we'll take Dr. Crusher's word for it that the syndrome is the result of a point mutation (the substitution of a single base pair) in the genome of an adult. She mutters in an aside that neither parent had the mutation; this bothers her, but then she knows that spontaneous mutations do occur. In fact, this circumstance is consistent with many known genetic diseases. While most genetic diseases are inherited from one or both parents and have been passed on through the generations—sickle-cell anemia is an example—some are caused by point mutations that occur spontaneously in the embryonic genome. More rarely, mutations in the mature adult can cause disease. Most cancers, for example, are probably started by a point mutation in an adult cell.

Now Dr. Crusher does something unexpected, at least by current medical standards. Once she suspected Forrester-Trent syndrome, she should have assayed for the genetic mutation that caused it, but she doesn't do that. Vigo told her that the first signs of his disease had developed only in the past few weeks. Crusher may have suspected that because Vigo had developed one point-mutation illness as an adult he was at risk for others, or maybe she just wanted to be thorough, since Vigo was Picard's son. At any rate, she skipped the specific assay for FTS and went straight to the radical procedure of sequencing Jason's entire genome. (We'd like to know how she got *that* past the Starfleet HMO's utilization review commission!)

And of course once she had the complete DNA sequence in hand, she discovered what Bok had done. We think she may have sequenced the DNA from some cell other than one of Vigo's doctored white blood cells. Perhaps she used the DNA from one of his neurons, since Forrester-Trent syndrome is a neurologic disorder, and discovered not just the absence of any of Picard's characteristic polymorphisms but the puzzling fact that Vigo was chimeric—in essence, two different people. While his resequenced white blood cells showed one set of genes, the other tissues of his body had a different genetic sequence, as if they belonged to someone else. Then she may have obtained Picard's complete genome and asked the computer to run a comparison between it and Vigo's. In this case, she would have discovered that the 50-percent similarity she expected to see was not there. Plot foiled!

Vigo, with his disease in remission, has elected to return to Camor V, and as Dr. Crusher says good-bye to him he thanks her warmly for her help. This is a different person from the restless disaffected young man they had beamed aboard the Enterprise *a few days ago. He's a changed man, she muses to herself and then laughs. That's the literal truth, she thinks. Of course he's changed—he's got 0.01 percent of Jean-Luc's DNA in his cells!*

Can it be that this is enough DNA to change a man? Could the resequenced DNA have included genes for Picard's personality traits? Not if the changes were confined to white blood cells, of course. The relationship between genome and personality is far from clear, but there is compelling evidence for genetic inheritance of personality traits. For example, dopamine is a neurotransmitter that is crucial to the attention system of the brain (among several other functions). A particular gene that codes for a dopamine-receptor site in the brain has been associated with the character trait of risk-taking behavior. If the dopamine-receptor gene was located close enough to a polymorphism that Bok's resequencing was aiming for, and if the brain neurons were resequenced, Vigo's personality might have been modified. You can never be quite sure what will happen when you go messing around with somebody's genes.

EUGENICS, PRO AND CON

Given all the advanced expertise in genetics that must have been realized by the twenty-fourth century, it is interesting to note that genetically-enhanced individuals are not eligible for service in

Starfleet. In fact, genetic enhancement is illegal throughout the Federation. Which brings us to Julian Bashir.

Dr. Bashir graduated second in his medical school class. He often tells the story of how he missed first place by the simple error of mistaking a preganglionic fiber for a postganglionic nerve—but perhaps it was just as well that he ended up in second place. He couldn't afford to be too smart. Dr. Bashir also loves racquetball. His competitive nature, quick thinking, athleticism, and excellent coordination make him a natural. But in racquetball, as in life, he had carefully schooled himself not to be too good. He would win the lower rounds of competitions, make a few judicious slips, and seed himself so that he had the chance to play a lot of raquetball without drawing attention to himself by winning tournaments. He had to be careful. Why? In *DS9:* "Dr. Bashir, I Presume?" we learn that if anyone had found out that he was genetically enhanced, it would have been the end of Bashir's career.

The Federation injunction against genetic enhancement dates from the time of the Eugenics Wars on Earth (interestingly, these wars took place in the mid–1990s, according to the *Star Trek* timeline), when genetic researchers, eager to demonstrate the utility of their work to skeptical governments, moved ahead with breeding experiments, creating a race of violent and ambitious "Supermen." The genetically enhanced children of the Darwin Genetic Research Station (*TNG:* "Unnatural Selection") provide another example of what can go wrong when you mess with the genome: their enhanced immune system made them lethal to normal humans (more about this in chapter 7).

And in fact medical researchers of the real 1990s are attempting to develop techniques of DNA manipulation that will enhance the lives of some sick people. The hope is that physicians will soon be able to cure illnesses that result from simple mutations in the genome—so-called single-gene diseases, like cystic fibrosis and sickle-cell anemia. If the mutant gene is replaced by a healthy copy in the right tissues of the body, the person will no longer have the disease. Other types of inbred weaknesses, like the tendency in some breeds of dogs to form cataracts, may also prove to be amenable to treatment through genetic manipulation.

There are several necessary steps in this type of therapy, and they are by no means easy. First, you have to manufacture a correct copy of the healthy gene. Then you have to devise a delivery system, or vector, for the new gene in order to get it into the right cells of your patient. Then you must be sure that the new gene will be inserted into the right site on the chromosome. And then you have to ensure that the new gene will "turn on" at the right time, make the right amount of its product, and then turn itself off.

In the case of cystic fibrosis, the damaged gene is known. Scientists have labeled it *cftr*, for "cystic fibrosis transmembrane conductance regulator." Cystic fibrosis (CF) is caused by a mutation that results in the production of a thick mucus in the lungs and other organs of the body. In the lungs, the secretions accumulate, causing breathing problems and predisposing the person to recurrent pneumonias. CF patients also often have trouble digesting their food, because the thick mucus in the pancreas blocks the release of needed digestive enzymes into the gut. These enzymes are fairly easy to replace; doctors prescribe enzyme capsules to be taken with the patient's food. But clearing the thick, gummy mucus out of the lungs is difficult. Eventually the mucus clogs the lung sacs altogether, and without vigorous therapy CF patients die in late childhood or as young adults. CF researchers are looking at various respiratory viruses as possible vectors to carry a corrected *cftr* gene into the lung sacs—right where it's needed. The researchers have one advantage here. All they have to do is get one working copy of the gene into the genome of the cells lining the lungs. The new gene can go pretty much anywhere in the genome, as long as it doesn't interfere with some other gene activity vital to the lung cells; if all goes as planned, it will function well enough to produce liquid mucus that can clear the lung sacs.

Julian Bashir was born with serious learning disabilities. While he was still quite young, his parents took him to a non-Federation planet so that he could undergo the resequencing of his flawed genes. The resequencing greatly increased not just his intelligence but also his eye-hand coordination. In his case, the line between genetic therapy and genetic enhancement is fuzzy. Replacing a defective gene is one thing; messing with a healthy genome in an attempt to bring about an enhancement of function is another.

Would the restoration of "normal" brain organization to a child who had learning disabilities be considered genetic therapy or genetic enhancement? What if the child suffered from a mild mental retardation? How much developmental delay would have to be present to justify genetic intervention? As long as you are intervening in the genome of a child to compensate for a developmental delay, would not some parents say, "Go ahead and give them a few talents, too"? Where do you draw the line? Current physicians face similar dilemmas. For example, endocrinologists administer growth hormone to children who are deficient in this hormone and would otherwise end up significantly shorter than normal. They do not administer growth hormone to healthy children whose parents hope they will star in the NBA. But how short is "too short"? How tall is "normal"?

If the *Deep Space Nine* episode raises questions of ethics, the technological underpinnings are even more dubious. Intelligence and eye-hand coordination result from the interaction of many genes, over time. Both intelligence and eye-hand coordination are strongly influenced by practice and by appropriate cultural and environmental stimulation. It will be very difficult even to determine which genes are responsible for these traits, let alone what particular resequencing changes to introduce in order to "correct" them. Research is currently under way at several institutions to associate specific polymorphisms with multifactorial traits, like intelligence, so that at least we will know which regions of which chromosomes to study.

One of the main tools used to figure out what specific genes do is a laboratory creation known as a knockout mouse. In these mice, both of the parental copies of the gene whose effects are to be studied are damaged or eliminated, by means of genetic engineering, in hopes that the defect will be expressed in the animal. The engineering works like this. A copy of a disrupted (that is, mutated) gene is inserted into a stem-cell tissue culture. Some of the stem cells take up the disrupted gene in the right location, by homologous recombination, and these cells are then injected into a live mouse embryo that hasn't yet undergone cell differentiation. When the mouse grows up (assuming it survives the procedure and the genetic mutation), it is a chimera—a blend of two different genomes. Researchers are careful to use mouse strains with coats of different color, so that they will be able to tell which mouse embryos incorporated the disrupted gene—the chimeric mice look like calico cats. The next step is to breed a chimeric mouse to a normal mouse. A few of the offspring of this mating will have one copy of the normal gene from the normal parent and one copy of the disrupted gene from the chimeric parent—that is, they will be heterozygous for the disrupted gene; these offspring, too, will be identifiable by coat color. The final step is to breed two of these heterozygous mice to each other. One-quarter of the offspring of this mating will have both copies of the normal gene, one-half will have a normal copy and a disrupted copy (just like their heterozygous parents), and one-quarter will have inherited two copies of the nonfunctioning gene, so that whatever the gene was supposed to code for will be missing in the mice. These last are the knockout mice, and they will show the effects of the missing gene.

Sometimes the missing gene turns out to code for a vital process and none of the knockout mice survive. Sometimes the gene turns out to have a backup system and the mice therefore appear normal despite its absence. Sometimes the effects will show up at a particular period in the animal's development, or only when the animal is exposed to a specific drug—say, one that activates a particular enzyme system. Strains of knockout mice allow scientists to home in on very specific effects of various genes. Knockouts might be the key to identifying genes that shape intelligence, mathematical skills, personality traits, and other types of complex behaviors.

Dr. Bashir was visited on Deep Space 9 by four other Terrans who had undergone the same genetic enhancement procedure that he had as a child. These four were all geniuses, but they had never put their abilities to use. In fact, they were almost unable to cope with normal life. Their therapist hoped that meeting with Julian Bashir and seeing that one of their number had succeeded in the unenhanced world would inspire them to try to fit into society. Unhappily, Bashir's visitors showed significant personality impairment—most likely as a result of their genetic "enhancement." There's truth in this cautionary tale. Human intelligence is a mixture of verbal/language skills, nonverbal mathematical and spatial processing, associative ability, memory, and visual and auditory processing. But there are many other factors that contribute to an individual's optimum development—motivation, social skills, and high moral character, for example. The fields of developmental psychology and neuropsychiatry are concerned with constructing theoretical models of the phenomenon of intelligence; most advances are made by studying people with selective deficits of brain function—for example, children with inherited learning disabilities or adults who show specific deficits following stroke or head injury. While the knowledge of brain function would fill a book, we can say simply that intelligence alone does not guarantee success in life. One must also have social skills—emotional intelligence, if you will—in order to contribute to society.

There is a way to work toward enhancing intelligence which

may entail fewer risks than direct manipulation of the genome, however. Khan and the other Supermen of the Eugenics Wars of the 1990s (how far in the future that seemed in 1967, when *TOS:* "Space Seed" first aired!) were the product of selective breeding rather than of DNA manipulation. Selective breeding has of course been used among domesticated animals and plants for centuries to enhance traits considered valuable. Dogs, for example, have been domesticated from wolves at three different times in Earth history. Although wolves and dogs are now separate species, some breeds of dogs can mate with wolves and produce fertile young. No DNA manipulation was involved in the development of the Chihuahua or the poodle, only generations of selective breeding. A Khan might well be developed through selective breeding with enough time and generations, but not (so far, at least) by any geneticists we know.

According to Mr. Spock, the downfall of the Earth scientists of the 1990s was that they forgot that "superior ability breeds superior ambition." Their Supermen went on to become dictators and plunged the Earth into a series of wars. While we don't necessarily subscribe to Spock's observation, we do acknowledge the wisdom of the Federation's stand against the artificially produced genetic enhancement of human beings.

The Human Genome Project will place within our hands, and within our lifetimes, the knowledge of how we are made. The science is there. Will we have the wisdom to know how to use it? It is our hope that the general public will be educated about the science involved—not so they know how to make knockout mice but so that they can make wise decisions. Few people really understand how nuclear weapons or nuclear power plants work, but enough of us understand enough to debate thoughtfully (more or less) about how and when to use these technologies. Human cloning research is another area of science that frightens many people. Legislators at both the national and state levels are considering bans and moratoriums to stop human cloning research. But cloning research offers the potential of an unlimited supply of transplant organs—by cloning the person's own healthy cells and growing a new liver, for example. And cloning research is going on all the time, in some unsuspected places.

DOLLY AND HER CHILDREN

Cloning is a natural process for many organisms. Your grand-mother encouraged cloning on her windowsill, whenever she took a leaf from a favorite violet and stuck it in soil to grow a new plant. Dandelions reproduce exclusively by cloning and most of us don't find dandelions threatening (or maybe we do—depends on your neighbors).

Dolly has become the poster clone of the nineties. "Little lamb, who made thee?" indeed! The answer is Dr. Ian Wilmut and his colleagues at the Roslin Institute, outside Edinburgh, who announced early last year that they had taken the nucleus of an udder cell from an adult "parent"—a six-year-old ewe—and placed it into the cytoplasm of an unfertilized egg cell. As we noted in chapter 1, the true scientific importance of this experiment was in the demonstration (which needs to be repeated before the results can be accepted) that the genome in the nucleus of an adult cell can still function as a complete set of directions for the development of an embryo into a healthy animal. Dolly was not the first animal ever to be cloned, but she was the first animal said to be cloned from an adult cell.

Before Dolly, the word "cloning" in the popular parlance had come to refer to the creation of an identical adult organism from another adult through technological intervention. This king-size blooper was committed by the writers of TNG: "Up the Long Ladder." Early in the twenty-second century, the S.S. Mariposa, on a colonizing mission, crash-landed on a planet in the Ficus Sector, and only five colonists survived the crash. To create a viable population, the Mariposans were forced to turn to cloning. To augment their gene pool, the Mariposans kidnap and "clone" Commander Riker and Dr. Pulaski, thereby immediately producing two adults. But organisms produced by cloning, like Dolly, are embryos. They take as long to mature as do babies that result from ordinary sexual reproduction.

In *TNG*: "Second Chances," even the estimable Dr. Crusher thought at first that she was faced with Riker's "clone," when his exact double was discovered on Nervala IV, having been marooned there for eight years. What in fact had happened was that when Riker visited Nervala IV on the *Potemkin* eight years earlier, a transporter accident somehow "split" him into two complete incarnations of himself (this is an odd bit of biophysics that we won't go into). If, instead, Riker had been cloned on Nervala IV eight years ago, the clone would be an eight-year-old child, not someone who appeared to be Riker's identical twin.

When Dr. Crusher sequenced the genomes of the two Rikers, she found "near total identity." This surprised her, because she knew that cloning ought to produce a certain number of discrepancies, but she had found only one or two. She was at a loss to explain the virtually identical DNA sequences—until the eight-year-old transporter logs of the *Potemkin* were examined and the creation of the second Riker was explained. The error rate in transporter technology has to be much lower than the error rate in cloning, in order for the transporters to successfully reproduce people time after time, down to the exact chemical patterns of their brains that code for thoughts and memories.

But something else puzzled us about this episode—Commander Riker's instinctive dislike of his doppelgänger. Identical twins typically report the opposite sensation—they feel a close emotional bond. They enjoy and desire each other's company. One would expect that the two Rikers would have had a lot to talk about! If you discovered that your genetic duplicate had taken your place and had survived in isolation on an inhospitable planet for eight years, wouldn't you want to know how "you" had done? Wouldn't you be fascinated by someone who understood you intuitively, who could relate to all of your formative experiences, but had met and handled a different set of challenges?

The two Rikers shared something that true clones of adult organisms do not—a childhood. Since they were created as a result of a transporter beam accident, each was legitimately *the* William Thomas Riker who had existed up to that point. The nature vs. nurture debate increasingly centers around the consideration that we are more than our genes—we are the sum of our experiences as well. Consider once more the cloned society of Mariposans. It would likely function quite normally, although cloning might give rise to some interesting governmental structures. Representation might be considered adequate if one member of each clone tribe were present—one Alice, one Bob, a Carol, a Dennis, and an Elizabeth. But even clones are not simply the sum of their genes. Identical twins are clones—that is, they are genetically identical—but they are also individuals, shaped by their own ideas and experiences.

Philosophers, psychologists, and geneticists will continue to wrangle over the nature v. nurture debate, but one question that genetics research may find an answer to sooner—a question that many people find almost as compelling—is whether we can significantly extend the limits of our human life span. Threescore years and ten seemed ample in Biblical times, and probably seemed fine to you when you were a teenager. But the older we get, the less satisfactory it looks. In the next chapter, we examine what *Star Trek* has to say about the age-old problem of aging.

CHAPTER SEVEN

Twenty-nine Years and Counting

"You see, we are mortal. Our time in this universe is finite. That is one of the truths that all humans must learn."

—Picard to noncorporeal entity (TNG: "The Bonding")

Mc*Coy walked the halls of the* Enterprise-D *with mixed emotions. He owed a lot to the* Enterprise; *he had changed so much during his tour of duty. He had even learned to trust Vulcans. But this ship wasn't his* Enterprise. *He was glad to have this tour of Starfleet's flagship, but sorry to find that the* Enterprise-D *was so different from his* Enterprise, *which had burned in 2285 over the Genesis planet. It wasn't so much the ship he missed, but the people. Well, he mused, that's the price of getting old. You accumulate losses. After a while, your shields go down, and the next photon torpedo is for you.*

Admiral McCoy was white-haired, tremulous, slow, and a curmudgeon—still a curmudgeon. Data noted these aspects of human aging without comment—and also the long cartilage in McCoy's nose and ears, the lack of elasticity in the skin, and the loss of sub-

cutaneous fat, which gave the old doctor's face its deep character lines. If Data had taken measurements, he would have found that McCoy's bones were thinner and his muscle mass had decreased. But, then, McCoy was a hundred and thirty-seven years old. . . .

Hold it—a hundred and thirty-seven years old? Suddenly, to our twentieth (nearly twenty-first) century eyes, the McCoy of *TNG's* "The Encounter at Farpoint" seems pretty spry. We should all hope to be that fit at a hundred and thirty-seven!

In this chapter, we will examine aging and the question of whether or not the human life span can be lengthened to the extent suggested by the *Star Trek* writers. The current human life span is worthy of our attention, for that matter. Why do people live so long, relative to the other animals of planet Earth? What determines the average life span of a species? There are *Star Trek* episodes in which the plot centers around hyperaging and others in which the aging of an individual is slowed or altered in some way. What determines how fast we age? Is there any way to slow aging down, or stop it altogether? How about putting it into reverse?

LIFE CYCLES

In chapter 1, we discussed morphology and a bit of embryology— how tissues form from the fertilized egg and develop into separate organs. An organism is considered mature when it has reached its adult size and weight and is capable of reproduction. In many species, including our own, the capacity for reproduction precedes the achievement of full maturity (in the United States, the onset of puberty is currently at about eleven-and-a-half years of age), and adulthood can last for many years after maturity.

However, there are numerous earthly life-forms that live until they breed, and then decline and die. Several species of insect—the mayfly is one example—do not even have digestive systems in their adult form: death is programmed to follow reproduction. The mayfly has twenty-four hours after metamorphosis from its larval stage to find a mate and lay eggs. The cicada can spend as long as seventeen years in its larval phase. Once they emerge in adult form, the males sing loudly, they mate, the females lay eggs, and the whole generation dies within a week or two. Pacific salmon are another example. After up to five years in ocean waters, they reach a maturity that compels them to return to the river where they were spawned. Their physiology and appearance change markedly during the journey. Having arrived exhausted at their home stream, they breed and die. Atlantic salmon do not face this programmed senescence. If they survive the rugged upstream swim to breed, they return to the ocean to repeat the cycle in another year or two.

Animals and plants that die after reproduction usually have lots of offspring, since the young have to fend for themselves. A salmon that lays several hundred eggs can afford to leave them on their own. Even if hundreds of eggs are eaten, or most of the hatchlings die, the parent salmon will be a prolific producer if a dozen fingerlings survive to reproduce. Birds and mammals follow a different plan. They have very few babies, but they look after their offspring to ensure their survival. Birds and mammals have helpless babies that have to be fed, kept warm, clean, and safe from predators. Most are dependent on parents for learning fundamental living skills as well.

Humans, as it happens, endure the longest childhood of all mammalian species. This period of dependency guarantees that human children will get some training and experience in life before they have to face it on their own. Spending nearly two decades in the company of adults who were healthy enough to reproduce gives human children the opportunity to learn the skills needed for survival, whether that means knowing how to pilot a starship or hunt for their dinner in the forest. So, does our lengthy childhood determine our relatively long life span? Or does our long life span enable us to have a childhood that lasts such an unusually long time? Maybe if we compare humans with other humanoid species, we'll get an answer.

On one end of the life-span continuum are the Ocampa. As we noted in chapter 5, these extraordinary humanoids from the Delta Quadrant have an average life span of only nine years. Like short-lived Terran mammals, they reach adult size in only one to two years and sexual maturity between four and five years of age. This means that an Ocampan has to pack all the learning that human children acquire over sixteen or so years into three or four. Kes, the resident Ocampan on *Voyager,* surprised the Doctor by the speed with which she digested the contents of the medical books he gave her. She clearly had some way of rapidly assimilating information. Ocampans are capable of telepathy and telekinesis. They had better be. Even with an extraordinary maturation of the nervous system, it is doubtful that a humanoid could acquire within three or four years enough of a knowledge base to become a fully functioning adult in a complex society. Kes would be changing profoundly every few weeks, in order to pack into a short life of nine years all of the character development that a normal human undergoes in seventy years. This would be unsettling to the humans around her, who are accustomed to going through life's developmental stages together. Kes would start younger and more inexperienced, then become an equal, and then advance in knowledge and wisdom well beyond her former peers. Most humans wouldn't handle that very well.

At the other end of the scale are the El-Aurians. Individuals belonging to this humanoid species of "listeners" live for several hundred Earth years. They are a quiet and dignified people, who were scattered throughout the galaxy after their homeworld was destroyed by the Borg. Accordingly, they do not discuss their own culture very much, and they adapt readily to other cultures.

Guinan is the representative El-Aurian aboard the *Enterprise*-D. As we would expect of someone more than five hundred years old, she is wise. She is also highly intuitive. She has acquired character and depth. She doesn't offer opinions until she is asked, or unless she recognizes that the right word at the right time might help to put someone on the proper track. She's had a number of children, but only one son seems to have given her any trouble—in "Evolution," she tells Dr. Crusher that he "wouldn't listen." However, after seventy years or so he straightened out—which suggests that El-Aurian children endure a lengthy adolescence, commensurate with their extended life span. Her colleagues aboard the *Enterprise*-D see Guinan as the epitome of stability; in their eyes, she is changeless.

The extreme difference in life span between the Ocampa and the El-Aurians leads one to speculate about what their societies would be like. When would citizens of each society become eligible to vote? To drink? How long would they be required to stay in school? How old would they have to be before they could marry without parental consent? When might they be slated for retirement? If a lot of El-Aurians decided to take up careers in Starfleet, they would eventually hold all the positions of command; no matter how long a human being served in Starfleet, there would likely always be an El-Aurian who had seniority. On the other hand, an Ocampan would be lucky even to graduate from Starfleet Academy, since the usual four-year program would take nearly half of his or her lifetime.

As humans familiar only with the fauna of Earth, we find it strange to contemplate such extreme differences in life span in humanoids. When we think of animals with short life spans, we think of small rodents, like mice and hamsters, who tend to live only one to three years—and then there are those tiny twenty-four-hour mayflies! On Earth, the larger animals seem to be the long-lived ones. However, while there is an association between body size and metabolic rate and an association between metabolic rate and life span, the relationship is not straightforward. For one thing, the average life span of an animal species is not an easy thing to determine accurately. Many animals live much longer as pets or in zoos than they do in the wild—but what is their species' average life span? Do you count the average number of years the animal survives in its wild habitat? Or do you factor in the life span of the oldest representative of the species, who lives in a zoo? Sometimes you find that measures of life span are actually measures of how hard it is to survive.

Improvement in our species' ability to survive probably explains the advance in human life span over the past 100 generations or so. The citizens of ancient Rome, on average, lived only into their mid-twenties. Americans born around 1900 had a life expectancy of between forty and fifty years. Babies born today in the United States can expect to live to be age seventy-two if they are male and seventy-nine if they are female. And so it seems quite reasonable to suppose that by the twenty-fourth century humans will be living until they are about a hundred and twenty years old or so—right?

Not so fast. The pattern of lengthening human life span is there, but the reason is not that we are winning the battle against the aging of the body. Rather, we are winning the battle of survival. We have managed to stop many diseases that used to kill our newborns, and infant mortality has its statistical effect upon life-expectancy rates. We have found ways to give our populations better food, better water, better shelter from the elements. But even in ancient Rome there were senators who lived into their sixties. The Bible tells us, in Psalm 90, that "the days of our years are threescore years and ten." Like our earlier problem of comparing the life span of wild animals with that of their counterparts in zoos, the average life span of a population of humans at some point in time is not an accurate measure of how old humans can get to be under optimum conditions. Even now, the expected life span in some Third World countries is only the mid-forties—not because the inhabitants cannot live as long as people in developed countries but because conditions are so harsh that they don't.

WHY WE GROW OLD

So let's go back to the metabolism question. Should Federation scientists be trying to make humans larger in order to grant them longer lives? Ater all, horses live longer than hamsters. And don't horses have a slower metabolism? Should the Federation try to slow us all down, too? These questions have puzzled naturalists for generations, but the answer to both is probably not.

What at first appears to be a life-span advantage for large animals is probably a survival advantage. Small creatures lose heat more rapidly, are more threatened by environmental catastrophes (in a flood, would you rather be a hamster or a horse?), and are more often eaten by large creatures. It also turns out that hamsters and horses have similar metabolic rates. Metabolism, when measured at the tissue level, is about the same for all mammals (except in special circumstances like hibernation). Whether you are a hamster or a horse, cell division takes about the same amount of time, and the ATP reaction releases the same amount of energy. It is true that hamster hearts beat faster than horse hearts; this is because hamster hearts are smaller—watches tick faster than grandfather clocks, but they mark the same time. Since hearts are biological and have only so many ticks in them (about 800 million for the life span of most mammals), hamster hearts wear out in fewer years.

The relationship between size and longevity becomes more complex if we look within a species. It turns out that dogs of small breeds live longer than dog of large breeds. Size, by itself, does not predict life span.

But as with so many rules in biology, it is the exception that becomes the most instructive. Songbirds, such as sparrows and finches, are no bigger than hamsters, have hearts that beat quite fast, but live longer than most rodents. In all likelihood, differences in life span among animal species are determined by the reproductive cycle. Once again, size is a consideration, but only a secondary one. It takes longer for a big animal to reach its full size. Horses take eleven months to develop in the womb, and it takes three years for a horse to reach sexual maturity. Moreover, partly because of their large size, horses can give birth to only one offspring a year. Hamsters are considerably quicker—gestation is only fifteen days—and they raise six to twelve (or more!) pups in each brood. It takes only eleven weeks for baby hamsters to reach maturity and have their own pups. Individual hamsters don't have to live very long in order to ensure the survival of the species. Sparrows, however, reproduce more slowly. Baby sparrows, while able to fly and fend for themselves within a month or two, are not sexually mature until the following year. A pair of sparrows can raise three broods in a summer, if they hurry and are blessed with good weather and few predators. Life span seems to correlate most with reproductive success—the longer it takes you to bear and raise children, the longer you live.

You're also granted extra years by the evolutionary process if you help your grandchildren. Elephants live a long time compared to other social animals, such as wolves and zebras—not because they are bigger but, some evolutionary biologists believe, because of their child-rearing habits. Elephant herds are led by matriarch elephants, whose lifetime of wisdom contributes to the survival of the herd's grandchildren and great-grandchildren. There is an evolutionary advantage for elephants to live to great age, past the point of active reproduction.

So what about us? If we want to "live long and prosper," what do we have to do? Well, for the "prosper" part you might want to

consult an accountant, but the "living long" part is more compli-
cated. After our discussion of natural selection, you might think
that the whole story is determined by genes; however, it turns out
that genes are far from being the only consideration. Studies of
twins—both fraternal and identical twins and twins who were
raised apart and raised together—form the basis of our knowledge
about which characteristics arise from heredity and which from
the environment. By combining the statistics from hundreds of
these twin pairs and running a mathematical analysis of variance,
scientists have shown that most of the determinants of life span in
humans are external factors of lifestyle and environment. Only
about a third of the factors responsible for determining the life
span of an individual are intrinsic—that is, determined by the biology of the individual.

We perceive aging as a single process, but in fact aging results
from several things happening all at the same time. We become
stooped and shorter and sometimes acquire a "dowager's hump"
because our aging bones don't deposit calcium as quickly as it is
lost. Tiny compression fractures in the vertebrae make us shorter.
Wrinkling and spotting of the skin and graying of the hair are the
most obvious changes of aging. Hair turns gray because of the
death of melanocytes—cells in the hair follicles that make hair
dark. Wrinkles are principally caused by what is known as the
cross-linking of collagen proteins: molecules that have lain side by
side for years begin to fuse chemically, with the result that the skin
is not as elastic as it once was. Now, when it stretches, you start to
get little breaks in the protein substructure. Subcutaneous fat layers are lost. On the outside, you start to look wrinkled.

Age-related changes in the brain—memory loss and slower
recall—are caused by the cumulative mortality of the nerve cells,
or neurons, over the years. When the cells in many of our other
organs die, they are replaced. Not so with neurons. The brain contains its full complement of neurons by the third trimester of preg-

nancy. As noted in chapter 5, many will be pruned in the first three years of life, as the neural pathways are established, but no more neurons will be produced. The dementia that arises from Alzheimer's disease—a condition that was once referred to simply (and unhelpfully) as senility (from the Latin *senex,* for "old" and "old man")—is probably caused by the accumulation of amyloid and junk proteins in the brain cortex, leading to neuronal losses. Memory loss and other signs of mental "slowing down" can also be caused by small strokes. High blood pressure can cause bleeding into the brain; atherosclerosis can plug the blood vessels. If you've lived a life of "hard knocks," there may be physical injury as well. All such damage can lead to neuronal loss.

In other words, age is the result of many different things going wrong. We usually see them all happening together, so we tend to think of aging as a singular condition, gradually approaching, like a rainstorm on the horizon. We all recognize a clunker car on the road when we see one, but it got that way because of rust on the undercarriage, corrosion in the engine, wear on the bearings, and a few accidents along the road of life.

Some of those accidents are not avoidable. In *TOS:* "The Deadly Years," all the inhabitants of a Federation colony named Gamma Hydra IV, none of them over thirty, die of some mysterious disease that looks like old age. Naturally, the *Enterprise* investigates, and in one of the most striking episodes of the original series, the members of the landing party "come down with" a rapidly advancing case of aging after they have returned to the ship. In a matter of days, Kirk, Scotty, McCoy, Spock, and Lieutenant Galway are all gray, wrinkled, palsied, and getting worse by the minute. Kirk, growing senile, is relieved of command after a humiliating competency hearing. In a race with his own death, the trembling McCoy formulates a remedy, based on the observation that Chekov, alone among the landing party, did not age. The damage turns out to have been caused by the recent passage of

Gamma Hydra IV through the tail of a comet that was streaming radiation in its wake.[1]

The episode doesn't specify the type of radiation emitted by the comet, and that would have made a difference. Various kinds of radiation leave specific signatures of damage on the human body. Ultraviolet radiation is hard on the skin—it disrupts the base-pair bonds of DNA, and UV light produces free radicals. Free radicals (in the chemical, not the political, sense) are usually oxygen with naked electrons ready for bonding. These reactive oxygen atoms insert themselves into any available molecule, and if it happens to be a skin protein or DNA, that's just too bad. X-ray radiation is more likely to disrupt rapidly growing cells throughout the body, like the tissue cells lining the gut and stem cells in the bone marrow. And remember those GCRs from chapter 2? Galactic cosmic rays do their share of damage to body tissues on Earth, although we are largely shielded from GCRs by the Earth's atmosphere. Like X rays, they damage DNA and so cause the most problems in rapidly growing tissues. But no type of radiation causes the hyperaging experienced by Kirk and the other members of the landing party. Fortunately for us Terrans, we have evolved enzymes that seek out and repair a lot of typical radiation damage. (Sometimes the repair mechanisms can't keep up, and ultimately, UV and other radiation exposure takes its toll on our bodies.)

In *TNG:* "Unnatural Selection," Dr. Pulaski encounters a different cause of hyperaging. In her case, the first symptom is arthritis in her elbow. Then, along with the staff of the Darwin Genetic Research Station, she finds herself rapidly developing wrinkled skin, white hair, and stiff joints and movements. To outward appearances, she seems to be aging with extreme rapidity. She finally concludes

---

[1] We realize that comets don't usually stream radiation in their wake, but we'll leave that problem to astronomers and stick with the biological premise here.

that the "hyperaging" is caused by an antibody attack by the incredibly proactive immune systems of the genetically engineered children of Darwin station. The genetic engineer explains that the children's immune systems don't wait to encounter infectious agents in the children's bodies—they reach out and attack a foreign-body threat, and the attack is destroying the adults' tissues.

Well, the *Star Trek* writers are half right. Antibodies do attack healthy tissues of the body sometimes, and can cause arthritis, renal failure, skin problems, glandular failure, and a variety of other ills. But the antibodies themselves don't destroy the tissues. What they do is mark the cells that are to be destroyed; it is the plasma complement system and the white blood cells that actually do the killing. Loose antibodies by themselves don't cause tissue damage (and, we might add, antibodies are seldom on the loose). Moreover, the kind of arthritis caused by aging is osteoarthritis (wear and tear on the joints), not an immune-mediated arthritis like rheumatoid arthritis or systemic lupus erythematosis. Dr. Pulaski's sore elbow, if caused by genetically engineered flying antibodies, would not have presaged hyperaging.

There are in fact genuine cases of accelerated aging—a cluster of diseases grouped under the medical umbrella term *progeria*, or premature senility. The human genome is subject to Murphy's Law: anything that can go wrong will go wrong. There are people who lack the enzymatic repair mechanisms for their DNA. Before they have fully matured—almost from the moment of birth—they begin to grow old. While they are still young children, they develop wrinkled skin and a wizened facial expression. They begin losing their hair, and they develop the thin limbs and the characteristically long nose and earlobes of old people. They are bothered by osteoarthritis and poor sleep. They are not old people—they are children—but, tragically, they die young, of old age. Fewer than 100 children are born with this affliction annually in the United States, and very little can be done to retard its progress. Genetic replacement therapy holds out real hope of a cure for progeria, since the defective enzymes have been identified and result from point mutations in the genome.

A TIME LIMIT FOR CELLS

Instead of asking why we age, let's ask how we manage to stave off age for so long. Considering that we are a collection of physiological mechanisms in constant use (your heart never stops beating, your lungs never stop breathing, your brain never stops running the show) until we die, it is a wonder that many more of our systems don't wear out sooner than they do. In fact, the body is equipped with mechanisms for ongoing maintenance and upkeep. Lots of organs are continuously being replaced: when old blood cells get banged up too much, they're taken out of circulation (literally) and replaced with new ones. But every time a new cell or protein replicates, there's another chance for a mutation in its DNA. The body attempts to exert some quality control in this regard by means of "proofreading" enzymes, which check for errors in transcription and signal for repair enzymes to fix the problem. Other enzymes are responsible for mopping up damage to the genome after exposure to ultraviolet or other radiation.

All this suggests that if you grow cells in a benign tissue culture, far away from any nasty radiation and (by virtue of their very isolation) spared the mechanical wear and tear of a career in the human body, they will go on dividing indefinitely. And indeed, among biologists this was the dogma for a long time, until an upstart graduate student in Texas named Leonard Hayflick actually tested the belief. He found that human skin fibroblasts in culture are capable of between fifty and seventy generations, and then they stop dividing and die, no matter how perfect the culture conditions are. When he tested cells from other animal species, he found that there were limits to the number of times they could divide, too—and that cells from long-lived animal species produced more generations than did cells from short-lived animal species. The phenomenon has become known as the Hayflick limit: no matter how you carefully coddle them, cells have a built-in obsolescence.

The search was on for the cause of this limit of survival. Scientists now believe they have found it. The polymerase enzymes that copy DNA during the cell-reproduction process (a.k.a. mitosis) can't get all the way out to the very last base pair of a chromosome. As when you sew on a button, there's always a little bit of thread that gets wasted when you tie off the knot. To take this mechanical problem into account, the ends of chromosomes are equipped with special sequences of DNA called telomeres. Every time a cell divides and the DNA replicates itself, a bit of the telomere is lost, and the chromosomes are therefore just that little bit shorter. (Since the telomeres themselves are not genes, no mutations result from this shortening.) After about seventy cell divisions, human chromosomes are all out of telomere sequence. Once the telomeres are gone, the cells stop dividing. Hayflick's observation is explained.

You may be wondering about gametes—eggs and sperm. You are the result of an egg cell that was formed when your mother was a fetus in your grandmother's womb. (Now there's something to ponder!) That was a long time ago—so how could the chromosomes in those cells have survived undamaged that long? It turns out that an enzyme called telomerase restores damaged telomeres. This enzyme is switched off in almost all normal cells, but it remains active in your egg and sperm cells to maintain the telomeres. Germ cells need telomerase—they can't afford to lose the capacity for unlimited regeneration, or we would die out as a species.

Recently Dr. Calvin Harley of the Geron Corporation in Menlo Park, California, and Drs. Jerry Shay and Woodring Wright at the

University of Texas Southwestern Medical Center in Dallas managed to increase the longevity of some human fibroblast cells by inserting telomerase into them. The cells have exceeded the Hayflick limit by several generations. Could this be the long-sought Fountain of Youth? Most biologists are keenly interested, but a bit skeptical at this point. The experiment proves that telomeres are directly related to cell senescence. It suggests that telomerase might be the key to coaxing damaged organs (heart, kidneys, or liver) to begin to grow healthy replacement tissues. The problem is to turn off the telomerase once it has done its job and the tissues have grown as much as they need to. Many cancers have found a way to activate telomerase. In fact, drug companies are researching antitelomerase agents for use in cancer chemotherapy.

There is a Greek myth about a goddess who took a handsome mortal as her consort. She granted him immortality but forgot to grant him eternal youth. He became increasingly frail and senescent, eventually turning into a cricket, which the goddess kept near her in a golden cage. In some episodes the *Star Trek* writers seem to have forgotten about the fact of aging, too. The long-lived humanoids we encounter on *Star Trek* must have *some* physiological way of maintaining their youthful appearance. There's Guinan, for one, and also the young and beautiful Acamarian Yuta of *TNG:* "The Vengeance Factor," who was able to pursue a century-long vendetta against the Lornaks. If Yuta had a gene for some sort of supertelomerase added to her genome, her cells might have been able to keep on dividing indefinitely. But her body would have been subject to a century's worth of wear and tear: skin, joints, and other systems would still be subject to all kinds of mechanical stress. Yuta was doubtless a fanatic about nutrition and exercise in order to keep herself going—a preoccupation that may have steered her toward her career as a chef.

The *Star Trek* record for longevity may well be held by Flint, who appears in *TOS:* "Requiem for Methuselah." Flint was born on Earth in the fourth millennium B.C. His tissues would instantly regenerate themselves whenever any damage occurred, which is how he managed to live long enough for Captain Kirk and the *Enterprise* crew to encounter him in the twenty-third century A.D. By the time of the encounter, on Holberg 917G, Flint was beginning gradually to die; apparently the "tissue regeneration" depended on environmental factors back on Earth. But for the greater part of his life, Flint presumably was able to beat both the telomeric Hayflick limit and the wear and tear of age. And presumably even his brain tissue "regenerated" itself, since he did not appear to be particularly demented in the episode. But Flint would have had very big feet. Some parts of us—the cartilage in our ears and nose, for example—keep expanding all our lives. Young beauties sport small, cute button noses. Old hags have long hooked noses. Flint would have had a whopper!

In the case of the long-lived children on the planet Miri, in the original series episode of that name, there was both a slowing of aging by a factor of 30 or more and a concomitant slowing of the maturation process. An "anti-aging virus" had escaped into the population from a research laboratory. The children remained in slowed aging until they reached puberty, at which time the virus responsible for their slowed aging produced a violent madness and death from necrosis of skin tissues. We can surmise that the viral agent acted on some type of DNA repair mechanism. To survive for 300 years, the children of Miri would have needed normal or better wound-healing, better than normal protection from collagen cross-linking and protein breakdown, and some way to keep from accumulating tissue where it wasn't needed.

We are unlikely to achieve immortality through engineering an anti-aging virus, but it is possible that the human life span can be lengthened to some extent by altering various extrinsic factors—accidents, nutrition, exercise, or disease. At present, the leading causes of death in industrialized nations—nations where famine, disease, and pestilence are not the limiting factors—are heart disease and cancer. When these illnesses are conquered, we will doubtless find ourselves confronted with other sorts of illnesses that do not crop up until one reaches the age of eighty or ninety. Assuming that medical researchers continue to work on extending the years of our productive life, the average human life span may reach a hundred years—or even a hundred and thirty years, as it is represented in *Star Trek*—but aging itself will still occur. Like beloved old cars, we may still be getting around, but our condition may depend on how well the owner performed the scheduled maintenance.

REVERSE AGING?

Picard, Guinan, Ensign Ro Laren, and Keiko Ishikawa discovered a kind of Beam of Youth as the result of a transporter accident in *TNG:* "Rascals." Oh, that transporter! McCoy was right not to trust the infernal thing. If it's not scattering your molecules all over the galaxy, it's creating clones, and if it's not cloning your evil twin, it's scrambling your DNA. In this case, the transporter masked the DNA expression of the four crewmen, taking them to an earlier point in their development, so that they resembled twelve-year-old children. The only thing is that their minds remained as they had been; somehow, the transporter buffer had scrambled their DNA enough to change its expression but hadn't done a thing to their brains. Come to think of it, their clothes came through the transport unchanged as well. Either the transporter correctly conveys a molecular pattern or it doesn't. You can't have corruption of the DNA molecules without making hash of a lot of other important biomolecules as well.

The twelve-year-old crew members weren't the first to undergo reverse aging. Admiral Mark Jameson wrangled an anti-aging treatment from an alien world in *TNG:* "Too Short a Season," initially hoping to treat his Iverson's disease, but then gave in to his lifelong ambition for work. If he could be young again, he could make a difference—maybe even make up for past mistakes. He was not destined for a happy second chance, however. His attempt to thwart fate and make up for past crimes ended in his death.

Reverse aging can't be a "simple" matter of reversing DNA switches to express genes that have been switched off—nor of going back to a previous state of DNA, when it had more potential. In human beings, reverse aging would have to entail absorbing tissues that had already fully developed. Your joints would have to get smaller; the cartilage in your ears and nose would have to be reabsorbed; cross-linked collagen fibrils would break their bonds and become more flexible; scars and injuries would disappear and tissues return to pristine condition; fat would disperse from central storage collections to subcutaneous regions throughout the body; bones and muscles would regenerate and become denser; some glands would shrink and others would reappear. No wonder Admiral Jameson was writhing in pain—he imploded! With reverse aging, we are quite firmly in the realm of fantasy.

The *Star Trek* writers have rightly decided that you just can't get around the fact of death. Death is on the other side even of reverse aging. We encounter a whole race of people who age in reverse in *VGR:* "Innocence"; they live on a planet in the Delta Quadrant called Drayan II. Drayans are born grown up. The episode dealt only with the end-of-life phase—unfortunately, since it would have been fascinating to witness the birth of a fully grown humanoid. The "oldest" members of the society return to their crysata (the sacred moon on which the race evolved) to die. When they have entered the final stage of their lives, Drayans look like children. They have small stature, youthful features, high-pitched voices, and are psychologically innocent and impulsive. This fools Tuvok into believing that he is dealing with children, and brings out his protective instincts and a little Vulcan discipline. Apparently, he makes all the right moves. The Drayan first prelate, Alcia, while incensed that the sanctity of the crysata has been violated by *Voyager's* incursion, recognizes a quality being when she meets one. Tuvok is allowed to accompany the Drayan he befriended through her final moments of life. Let us all hope that we have the company of such good friends on our final journey.

EIGHT

The Adventure Continues

"Bones! Are you afraid of the future?"
— *Kirk,* Star Trek VI: The Undiscovered Country

*A*fter Tom Paris's triumphant return from humanity's first voyage at warp 10 in the shuttlecraft, his DNA mutated. The Voyager *crew hadn't even known whether or not warp 10 was possible, let alone survivable—the theorists said that nothing could accelerate to warp 10, because at warp 10 you would be moving infinitely fast and therefore, theoretically, would be everywhere at once. However, desperate to find a way home, the crew had sunk all their energies into transwarp research. And now the mutated Paris had eloped in the shuttlecraft at warp 10 again, taking Captain Janeway with him. The crew tracked their signal as best they could, all the while terrified that both were lost forever. Eventually they traced the shuttlecraft to a Class-M planet, and the away team beamed down, hoping against hope to find Paris and Janeway—whatever they might have become—at least still alive. . . .*

On the planet's surface, the tricorder whirred and chirped with its customary efficiency, reassuring the team, as the signal grew stronger, that they had located the pair at last. Tuvok pulled the ferns aside and stopped short. Chakotay and the others gasped. They thought they were ready for anything, but they weren't ready for. . . . pink salamanders with puppies!

Are you ready for the future?

Star Trek has taken us to places no one has gone before— through the exploration of worlds possible and improbable. As we contemplate our own future, we might wonder whether we will do it the right way or mess it up. Sometimes we wish we could do things over and over again until we get them right, like the *Enterprise*-D crew at the Typhon Expanse in "Cause and Effect." The future of our civilization depends on our learning from the lessons of the past. As individuals, we get only one chance, no matter how much time medical science gives us for it. But as a species we do have plenty of chances to keep on trying to "get it right," as long as we don't self-destruct.

The concept of evolution is one of the great ideas of history. Our knowledge that the conditions of life are not fixed—that everything changes, and that there is always the potential for improvement as well as the possibility of extinction—has brought our individual mortality into sharp relief. Some people find the idea of evolution so threatening that they wage holy wars against it. Others find God's signature written in the marvels of the fossil record and the intricate traces of DNA. Whatever one's spiritual beliefs, the concept of evolution challenges us to recognize the biological heritage we share with all life on Earth. We can become more than what we are. We are part of something larger than ourselves.

GRADUALISM AND LEAPS

Evolution is usually thought of in terms of natural selection and speciation. Natural selection refers to the process by which an animal or plant species becomes better adapted to its environment: the strong and successful survive, reproduce, and pass their traits on to succeeding generations. Speciation occurs in various ways, commonly when a population is divided—as, for example, the squirrels were separated on the north and south rims of the Grand Canyon. Through the ensuing eons, their fitness was shaped in slightly different ways. There are now two different species of squirrels at the Grand Canyon, the Kaibab squirrel inhabiting the north rim and the Abert's squirrel on the south rim. The random mutations that occurred in one population did not occur in the other, and over time the two populations diverged. If there is intense pressure for survival, the changes that shape a population may occur quickly. If there is less intense pressure for survival, the basic model will remain intact for a long time. In the case of the Grand Canyon squirrels, it is thought that the populations diverged over a period of about 40 million years, roughly the amount of time it took the Colorado River to cut the canyon.

If the human population were to scatter across the galaxy, how much time would pass before evolutionary pressures began to distinguish one colony from another? Probably quite a while. Our cultural history as a species dates back only 5,000–10,000 years, but if we review what we know about the course of human evolution on Earth, we would have to look back about 2 million years—to the origins of *Homo erectus,* the hominid species that preceded us—before we saw an appreciable difference between ourselves and what we might term a primitive human. There would probably have to be a gap of at least 1 million or 2 million years between the humans at the turn of the twenty-first century and those of the future before natural selection would produce any appreciable differences. Some 2 million years would have to elapse, more or less (depending on environmental pressures and the incidence of mutations), before human colonies isolated in space would diverge into separate species. We cannot assume, however, that the timelines would all be similar. Survival pressures would doubtless vary, mutations would occur randomly, and therefore selection would advance at varying rates and different kinds of selections would be made.

There have been periods in Earth's history during which very rapid speciation has occurred, generally as a result of catastrophic upheavals in the environment resulting in the mass extinction of species and the consequent opening up of ecological "niches." The theory of periodic speciation is known as "punctuated equilibrium," and the classic paper illuminating it was written in 1972 by Stephen Jay Gould and Niles Eldredge. Gould has described for a popular audience some of the fascinating evidence for this story in his 1989 book *Wonderful Life,* which discusses the fossils of the Burgess Shale and the explosion of life-forms that marked the beginning of the Cambrian period 590 million years ago. (The idea of periods of rapid speciation was first proposed in the late 1940s by the Harvard paleontologist George Gaylord Simpson, who termed it "quantum evolution," but "punctuated equilibrium" as set forth by Gould and Eldredge has become the nomenclature of choice.)

Such a change occurred at the end of the Cretaceous period, a mere 65 million years ago, when the various species of dinosaurs and a large number of other species became extinct, thereby clearing the way for small mammals to proliferate and come to dominate the planet. The discontinuity in the fossil record at that point is striking evidence that speciation at certain times in geological history occurs on a broad scale and very rapidly. To understand what happened, let's regroup at what we recognize as the normal state of affairs.

Ordinarily, ecological balance creates relatively stable conditions. When a combination of animal and plant species find success in a certain geographic and climatic zone, those species tend to persist. Over time, they work out a relatively stable balance—so many rabbits and so many coyotes in the meadows, so many mosquitoes and loons on so many ponds, with a march of aspen and pines giving way to oaks and elms in the woods. Until some outside pressure occurs—for example, a climatic change due to advancing ice, or the introduction of an insect marauder, or the construction of a shopping complex—the ecological balance stays the same. Even when these changes occur, many of the same species will live on, having reached a different balance based on the new conditions. But what happens when an asteroid hits? Now you have a discontinuity!

An asteroid impact is the current best explanation for the mass extinctions at the end of the Cretaceous period. The evidence is in a layer of iridium (an element rare on Earth but often brought here by meteoric impact), which was demonstrated to be extraterrestrial in origin by Walter Alvarez of the University of California at Berkeley, with help from his father, the Nobel physicist Luis Alvarez. The iridium layer has since been discovered at more than fifty locations around the world, laid down exactly at what is known as the K-T boundary—the layer that separates fossils of the Cretaceous period from those of the succeeding Tertiary. More recently, what is thought to be ground zero has been identified: a depression about 170 kilometers across in the Yucatan Peninsula of Mexico; it is the remains of an impact crater, dating from the same time. The object that made the crater is thought to have been at least 10 kilometers in diameter, and the presence of the iridium "fallout" around the world suggests a planetwide disaster, probably in the form of a huge cloud of dust that blasted into the stratosphere and would have blocked light and warmth from the Sun. Creatures that weren't killed outright by the blast, the tidal waves, or the dust storms, endured months of darkness. Plants died; animals starved; the planet grew cold.

While this catastrophe is theoretical and has been reconstructed through circumstantial evidence, scientists have been able to measure the drift of ash and dust from natural catastrophes in our own time, like the the eruption of Krakatoa in 1883, the explosion of the Mount Saint Helens volcano in 1982, and the eruption of Mount Pinatubo in the Philippines in 1991. Ash from the Pinatubo and Mount Saint Helens eruptions darkened skies for miles and created spectacular sunsets a continent away for weeks afterward. The eruption of the Krakatoa volcano, one of the largest volcanic eruptions in recorded history, was heard over 2,000 miles away in Australia. Ash fell over an area of 300,000 square miles, and it took a year for the finer ash to settle out of the atmosphere. But the impact of the asteroid that made the Yucatan crater would have been greater by orders of magnitude. It would have triggered tsunamis and earthquakes all over the Earth, and the dust cloud would have darkened the daytime skies and lowered the Earth's temperature for months. Of course, with the possible exception of Q nobody was around to witness the event, but not having been around at the time has never kept scientists from speculating about things, the Big Bang being a prominent example.

Mass extinctions, however, are not the complete story. There are also periods of mass proliferation. Far earlier than the dinosaurs appeared, while most of the world was inhabited by simple organisms that were beginning to try out multicellularity as a lifestyle, something happened—an event that remains even more of a mystery than the Cretaceous-Tertiary boundary cataclysm. This event was productive rather than destructive. Near the end of the pre-Cambrian period, in a period of only 10 million years, evolutionary processes underwent a time warp. There was an unprecedented development in body plans. It is thought that the advent of an oxygen-rich atmosphere, along with global warming that released the planet from a long ice age, allowed life-forms to rapidly expand into unoccupied environmental niches. This might explain in part why new forms are found in such abundance at the start of the Cambrian, but the "how"of their rapid evolution is unknown.

There is at least one genetic process capable of producing radical changes in morphology by a single mutation. The genes involved are homeobox (Hox) genes. We touched on them briefly in chapter 1, when we discussed the possibility of humanoids growing antennae.

Hox genes are among the many genes that can be traced back to before the origin of vertebrates. Hox genes were first isolated in the geneticists' favorite experimental subject, the fruit fly (*Drosophila melanogaster*). Geneticists love fruit flies because they are cheap, small, reproduce rapidly, and don't bite. Fruit flies also have a number of natural mutant forms—variations in eye color and wing formation which have been studied carefully in breeding experiments to determine patterns of inheritance. As gene mapping became possible, *Drosophila* were employed in countless experiments to trace known genes to specific sites on the chromosomes. Nearly the entire genome of the fruit fly has been mapped and its genome is being sequenced in the Fruit Fly Genome Project. Researchers are now learning what happens when mutations are deliberately induced in specific *Drosophila* genes. Don't worry. You won't be reading about the fruit fly that ate Chicago. Genes don't have the ability to trigger the kinds of changes that transcend the laws of physics. But scientists have produced some pretty weird bugs (see figure). If you think that looks like legs growing where antennae ought to be, you're right.

That's what Hox genes can do—they determine which body segments grow in which part of the body. If you alter the order or number of Hox genes, it results in a change in the order or number of body parts. Hox genes have also caused flies to grow four wings

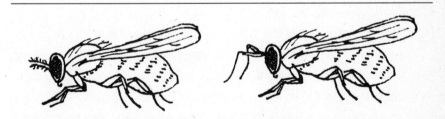

The effects of a Hox gene mutation in a fruit fly. The one on the left is a normal fruit fly. A Hox gene mutation in the one on the right has caused a pair of legs to emerge from the head.

rather than the normal two, or to grow no wings at all. In verte-brates, Hox genes can cause limb malformations. A specific muta-tion in a Hox gene in humans is responsible for some cases of polysyndactyly, a birth defect in which persons have fused fingers and toes and extra numbers of them.

Hox genes are among a few known families of genes capable of inducing major changes in body plans with a single mutation, and might explain the absence of intermediate forms in the fossil record. Such absences have been taken by creationists as "proof" that the theory of evolution is incorrect. For example, there are no fossil species of amphibious land mammals with tinier and tinier legs and larger and larger bodies, which would constitute some sort of morphological progression in the evolution of the whales. However, through the mutation of a Hox gene that controls limb or body-segment development, a wholesale change may well have occurred within a relatively few generations, producing a flipper from what had been a foot.

Such a change is possible, but not proven. Evolutionary theories are difficult to subject to experimental verification because of the long period of time over which evolutionary change takes place. The leap from cellular genetics to species evolution is an enormous one and highly speculative, but the connection is tantalizing and currently a source of intense research.

So, what about us? What changes might we see in ourselves? How will humanity evolve?

The best predicter of the future is the past, and if we look at the human past there isn't much to go by. *Homo sapiens* hasn't been around that long, compared with the history of life on Earth. Evi-dence points to the African continent as the site of human origins. *Homo erectus* rose there about 2 million years ago, and from there our ancestors migrated around the globe—north to Asia, then west into Europe and east to the far eastern reaches of Asia. Eventually, humans spread across Siberia and crossed what was then a land bridge into North America. They continued to wander, spreading to this continent's eastern seaboard and down into Central and South America. There is also evidence that oceanic voyages were undertaken, from South America to the Pacific islands and from northern Europe to North America.

During and following these migrations, over a period of a million years or so, we evolved all of the racial and ethnic diversity we now perceive in each other. Dark pigmentation conferred little advantage to people living in northern Europe, and it was eventually lost. African peoples remained relatively isolated from other continents and stayed heavily pigmented, but within the continent of Africa isolated subcultures developed different body habitus—like the short-statured !Kung San of the Kalahari Desert and the tall Tutsi of Rwanda and Burundi. On the continent of Australia, people developed the darkest pigmentation of all. Asian peoples are intermediate in coloring and share some facial features with their descendants, the native peoples of the Western Hemisphere. These differences emerged very gradually, over millennia of isolation. Now that our world is a smaller place, in which people of different continental origins mingle, we are likely to see these traits—pigmentation, facial features, body habitus, hair texture—blended, with eventual loss of distinct types. Beyond that, it is difficult to speculate about the physical changes the next couple of million years may bring. Such observable trends as our gradual increase in stature appear to be due more to lifestyle factors—better food, an easier existence—than to the forces of natural selection.

USE IT OR LOSE IT!

A cautionary tale we repeatedly encounter in the *Star Trek* universe is the evolution of the mind at the expense of the body. This condition is found throughout the *Star Trek* universe. Among the first examples we encountered were the Talosians—those big-headed, frail baldies who attempted to keep Captain Christopher Pike on Talos IV as the stud for their new servant race. The Talosians had enjoyed the life of the mind so much that they let their physical development go. They lived underground and vicariously, dependent on their virtual-reality technology to lead "full" lives. They became one-trick ponies, focusing on only one aspect of their humanoid existence to the neglect of all others. After selecting for more and better thought processes for generations, they found themselves unable to participate in the activity of life. Their race was dying. They began searching the galaxy for a servant race to work and play for them.

Star Trek revisits this theme with a twist in *TOS:* "Spock's Brain." The ancestors of the Eymorg of Sigma Draconis VI had made life so easy that their descendants (at least the ruling females, who lived below the planet's surface) forgot how to think or how to look after themselves. This loss of fitness is also seen in the Ocampa of the Delta Quadrant, another "kept" people. They were content to have all their needs provided by the overseeing Caretaker of their subterranean city, and so were unprepared to fend for themselves when the Caretaker died after 500 Ocampa generations.

The drive behind evolutionary improvement is the struggle for survival. You cannot afford to stop using either your mind or your body. The reptilian ancestors of snakes once had legs. The evolutionary moral seems to be use it or lose it!

WHAT THE FUTURE MAY HOLD—BUT PROBABLY WON'T

Most of us would find the idea of becoming mere *Homo cerebrus* repugnant. But what about the possibility of developing new cerebral powers? Can humankind evolve into a telepathic race?

Deanna Troi is a trained counselor, something like a therapist, but unlike a therapist she can dispense advice as well as insight. Sometimes her job is to tell it like it is to an angry superior officer, as she does when she forbids Riker to give in to his desire for grief and vengeance after it appears that Picard has been vaporized in a senseless barroom brawl (*TNG:* "Gambit"). More often, though, she uses her training and skills to bring the crew to a better appreciation of what is within them. As she does with the deaf negotiator, Riva, when he loses his translator chorus in "Loud as a Whisper," she encourages people to draw upon their inner resources.

As a half-Betazoid, Deanna is empathic. Telepathic races are relatively common among the peoples of *Star Trek*. Unlike "normal" telepaths, Vulcans require some physical contact in order to establish their mind meld, but then the connection is so intense that both parties share thoughts completely for the duration of the meld. Ocampans have prodigious mental powers, but do not develop them without training. Other races, like the Melkotians in *TOS:* "Spectre of the Gun," the Ullians of *TNG:* "Violations," and the Cairn of *TNG:* "Dark Page" are fully telepathic, using spoken language only to communicate with outworlders.

There doesn't seem to be much possibility of telepathy for humans, though. For all the stories that have been spun regarding extrasensory perception, Psi phenomena, and paranormal investigation, there has been no acceptable demonstration that telepathy exists. Nor have any scientifically verifiable demonstrations of telekinesis (the ability to manipulate objects through thought) been published. Granted, our subjective experience is extremely powerful. Sometimes we can feel as though the power of our mind constitutes a kind of energy. Certainly our thoughts, attitudes, and beliefs shape the experiences of our lives in ways that do not seem to submit to a rational accounting. What are we to make of basketball players who spend time visualizing themselves completing successful free-throws and thereby actually increase their percentage of successful shots? Or of golfers who "see" the shot before they hit it? Psychologists and therapists—even bartenders, like Guinan—continually encourage us to cultivate a positive attitude. Plants thrive for contented gardeners who talk to them. Is this evidence that thought has an energy of its own? The *Star Trek* writers seem to think so. In *TNG:* "Time's Arrow," we meet the Devidians, who live on "neural energy" that they steal from other life-forms. Onaya, a noncorporeal life-form, is similarly fueled by the thoughts of others—specifically their creative imaginings. To get what she needs, she will stimulate the production of neurotransmitters in your brain, enabling you to create great works of art but killing you off in the process. Apparently, one of her victims was the poet John Keats (*DS9:* "The Muse"). Energy feeders tend toward unpleasantness. Among the most unpleasant was the creature known as Redjac, who survived through the ages as a psychopathic murderer feeding on the "thought energy" of human terror (*TOS:* "Wolf in the Fold"). In *TNG:* "Gambit, Part II," the *Star Trek* writers come up with what must be regarded as the ultimate weapon—a "psionic resonator," with which Vulcans can focus their telepathic energy so that they can kill someone who is thinking violent thoughts.

Though it may sometimes seem to us that thought is a form of energy with the power to shape the world, the world that it actually shapes is our subjective world—the world between our ears. Our experience of the world is indeed altered by this internal experience, but not because of "thought energy"—there is no such thing. The missing explanation is an appreciation for the close integration of mind and body. The basketball player who visualizes herself successfully throwing the ball through the hoop is in fact rehearsing. The more carefully she imagines the process—imagines taking the preparatory dribble, mentally measuring the distance to the hoop, feeling the weight of the ball prior to launching it, continuing the follow-through while "watching" the ball go through the hoop—the more likely the visualization process will translate into "muscle memory" and increase the actual free-throw success rate. But the basketball player who spends *all* her time visualizing and no time on the court is not likely to score many points. Visualization works when it reinforces actual experience.

Another example of the close tie between mind and body is in the healing arts. Placebo controls are now routinely included in drug treatment trials. A placebo is an inert substance or treatment protocol designed to imitate the appearance and experience of the treatment under study—in every aspect except for the presence of the active ingredient. We are most accustomed to thinking of placebos as sugar pills, but in the case of some psychotherapeutic treatments, for example, the "placebo" may be an appointment that lasts as long as the therapeutic appointment but without the actual therapy. The occasional success of such placebo treatment is well-known, and certain types of illness are more likely than others to be influenced by the placebo. It is not uncommon for major depression to show placebo success rates of nearly 30 percent, and an antidepressant medication, like Pamelor or Prozac, has to demonstrate a statistically significant success rate greater than the placebo control before it can be marketed as an effective treatment.

Influences of the mind upon the body are real even if the mechanisms are not understood. Admitting that there is a mind-body connection we don't understand, however, is not the same thing as saying that the mind can do anything at all. Placebo response to treatment can be measured through standardized scientific protocols that satisfy the most skeptical of observers. Telepathy and telekinesis, on the other hand, have not been scientifically demonstrated. There is an enormous difference between affecting your own soma with your own brain and "sending" some sort of ineffable, metaphysical signal across the airways to the mind of another human being. As much as we might like the idea of telepathy—"My thoughts to your thoughts, my mind to your mind . . . ," we have no evidence that it exists, so there is no reason to expect that it will evolve.

I AIN'T GOT NO BODY

The future existence of humanity probably won't include development of the intellect at the expense of the body, or development of magical mental powers like telepathy. But how about reaching another plane of existence altogether? What about leaving the body behind and moving toward noncorporeal existence?

In *TNG:* "Transfigurations," a Zalkonian humanoid is found in critical condition, having barely survived the crash-landing of his shuttlecraft on a planet in the Zeta Gelis system. Dr. Crusher rehabilitates him, and is astonished to witness an ongoing physiological process that keeps him convulsed with pain. The crew likes the survivor; he is peaceable and seems to bear ill will toward no one—quite the contrary. Geordi finds that his acquaintance with the newcomer, whom the crew have nicknamed "John Doe," leads to an increase in his own self-confidence and skill with women. O'Brien's dislocated shoulder is instantaneously healed when John touches him. These minor mysteries are solved when an even larger mystery surfaces: John Doe mutates right out of the material world—much to the consternation of the Zalkonian police, who have been on his trail and are now helpless to apprehend him.

Noncorporeal species seem to pop up all over *Star Trek*. Besides Redjac and Onaya, there was the benign entity who decided to experience corporeal life as Deanna Troi's child, a welcome presence who most considerately left when it realized that it was placing the crew at risk by disrupting the stasis field surrounding some deadly plague virus cultures (*TNG:* "The Child"). The best known of *Star Trek's* noncorporeal life-forms, of course, is the Q continuum—although Q usually takes the form of a human male during his encounters with Picard, Sisko, and Janeway. And let us not forget Wesley Crusher and Kes. They both thought so hard that they thought themselves away. (Or they may be hovering nearby right now!)

There are two sorts of "noncorporeal" life-forms in the *Star Trek* universe. One is simply a life-form in some state of matter other than solid or liquid—hence we might have plasma beings or gaseous beings, who are considered noncorporeal because they do not have solid bodies. The other variety of noncorporeal life is an "energy being." For the most part, the *Star Trek* writers are sketchy on the details of noncorporeal life. It's pretty hard to conceive of a gaseous "object," alive or not, that could retain its coherence in our atmosphere and so be able to interact with Starfleet personnel. And coherence—the ability to code information, store it, retrieve it, and replicate it—is one of the necessary conditions for all life-forms. But if it's hard to conceive of a gaseous being, how about one that doesn't exist in the form of any sort of matter at all? Now we're out there on the fringes of reality again!

Energy does not form a pattern unless it is coded by some "thing"—and that means matter. Even if we conceive of a being that projects its thoughts (or its spirit, if you prefer) outward as an energy stream, the energy itself is not alive. It cannot interact unless it is received at the other end by some "thing." Nor can an energy stream modify itself; once it has been emitted, it can be modified only by some "thing." The hologram of Leonardo DaVinci (*VGR*: "Concerning Flight") was upset because he thought a woman was trapped in the *Voyager's* computer. We laugh, because we technologically more experienced humans understand that the voice of Majel Hall is not a woman at all but a synthesized (or recorded, depending on whether you're in the *Star Trek* universe or the real one) voice pattern. Likewise, the electronic signal that brings *Star Trek* to your television set isn't alive. It's a pattern generated by a "thing"—in this case, a transmitting station. At Paramount Studios, the cast itself is alive (or maybe you're watching syndicated reruns), but the energy you behold is not. Energy, by itself, cannot comprise a life-form.

So, if humanity is to evolve as a life-form, it will have to stay material. We aren't likely to become telepathic, which means we will continue to communicate, for better or worse, with words. We will remain in our own skins, essentially singular beings. Is there any hope for us to come together and reach the stars?

CULTURAL EVOLUTION

The most likely path of human evolution is manifest whenever Picard or Sisko or Janeway must explain themselves to a primitive and hitherto unknown alien race. How often have we heard them intone, "We were once like yourselves—savage, warlike, . . ." or words to that effect? In Gene Roddenberry's universe, the future according to *Star Trek,* the surest path for human evolution is cultural.

This is scientifically sensible. Cultural evolution is thought to account for our presence on Earth as the most numerous large species of mammal. Think about it. There are some 600,000 African elephants, 50,000 Asian elephants, 6,000 Siberian tigers, and only 1,000 or so giant pandas left in the wild. There are some 5 billion of us. We are crowding out a lot of other mammalian species. We are nearly naked, without fangs or claws, and not especially fast runners. How did we get to be so prevalent on the same planet that supports elephants, tigers, panda bears, and all the other threatened species? The answer is cultural evolution: we learn.

We may be, for the life-forms on our planet, the equivalent of the Borg threat. Any life-form—be it plant, animal, or prokaryote—that does not find a way to exist in some out-of-the-way spot is likely to be assimilated by us. Like the Borg, we are bent on perfecting ourselves. We take ideas and incorporate them into our body of knowledge. Also like the Borg, we can be transformed by an idea. Richard Dawkins has introduced the concept of memes. A meme is an idea whose dissemination transforms a culture—something like a paradigm shift. We see this happening in the *TNG* episodes "Descent" Parts I and II. Picard allows a captured Borg drone whom the crew has named "Hugh" to return to his Borg collective. Hugh carries back with him a destructive meme from the *Enterprise*-D: he introduces the concept of individuality, and his collective loses its Borg ability to function as a single mind. Hugh's collective is discovered by Lore as it drifts aimlessly through space, unable to take action because it can no longer focus and make decisions with its collective intelligence now divided into countless individuals. Like the Borg, the most remarkable changes we will undergo as a species are likely to come to us through our minds.

Human beings have opposable thumbs on flexible hands. We have a bipedal gait that frees our hands for something other than balance and locomotion. We use tools. We have a complex language. We have a frontal cortex that allows us to imagine our future and recall our past. And we seem to have an insatiable desire to explore. Once we have explored or discovered something new, we seem to have an equally insatiable desire to tell other human beings about it. We teach.

We have already experienced a series of cultural revolutions—fire, the wheel, agriculture, metallurgy, the printing press, the Industrial Revolution. Now we are embarked on three more—space exploration, the unlocking of the genome, and a communications revolution of unprecedented scale. We learn not just from direct experience but from the accumulated experience of other human beings. When we do not discard the learning of others through carelessness, warfare, or prejudice, we learn even faster. We have the potential to become better athletes and artists, to become wiser, better educated, more compassionate and tolerant—in a word, more "human."

That is the evolutionary future beckoning to us from *Star Trek:* to learn from our painful experiences of poverty, pollution, over-population, racism, and war. To learn that we need not be trapped by those evils forever or allow ourselves to be destroyed by them. To visualize what our future might be, make up our minds that we want to go there, and work to make it so.

CHAPTER
NINE

Where No One Will Ever Go

"It's a TV show. Get a life, people!"

— *William Shatner*, Saturday Night Live

It was a sublime experience. Picard knelt beside the puddle of scum that would form the first protein in Earth's history. On the horizon was the shimmering glory of a time-space distortion, brighter than the Sun. Even Q's cynical taunting couldn't completely tarnish the wonder of the moment. As he rose from the ground, dusted off his knees, and gave the customary tug to his tunic, Picard keeled over in a paroxysm of coughing, a froth of bloody mucus churning from his lungs. . . . (*TNG*: "All Good Things")

Well, no, it didn't exactly happen that way, but it should have. No DNA, so no plants. No plants, no oxygen. At this point in time, the Earth's atmosphere consisted chiefly of methane, ammonia, and hydrogen sulfide. Nasty!

Along with giving us characters who grow and develop into plausible people whom we care about, the *Star Trek* writers have taken the time, partly in response to pressure from their skeptical and fact-hungry fans, to try to make the series coherent and scientifically sound. *Star Trek* has always attempted to obey the laws of physics (unless it would interfere with a good story). Every now and then, though, you catch them with their scientific slip showing. Sometimes what appears to be a blooper becomes a chance to explore the boundaries of science and the mysteries of the "real world," as the preceding chapters have shown. Sometimes a blooper is just a blooper, and reveals a great deal about Hollywood and therefore something about our popular myths and values.

It's great fun to explore this wild side of the *Star Trek* universe. But hang on for the ride—we'll be traveling at warp speed.

SEX AGAIN

This blooper has haunted us for years.

Kirk, Spock, and McCoy watch with amazement as the electrical entity that so recently attacked them envelops Zefrem Cochrane. The mist of scintillating colors moves over the man, who becomes very still. The harsh yellows and oranges and static discharges give way to a soft glow of pastel roses and blues. Astounded, they realize that Cochrane is not a prisoner in this planet. The entity, whom Cochrane calls The Companion, isn't preying on him. It's in love. (TOS: "Metamorphosis")

As Kirk and Spock discuss their observations of Cochrane's Companion—that it is a "she" and deeply in love with the human Cochrane—Spock comments that male and female are universal principles for life throughout the galaxy.

As if!

By now, you realize that male and female aren't even stable conditions on Earth. Vertebrates show male and female sexual dimorphism, but many invertebrates have both male and female sexual organs—snails, for example. And many creatures that reproduce sexually, by exchanging DNA, are not of different sexes at all: remember the amoebae? And speaking of amoebae. . . .

LOOK! IT'S A BIRD! IT'S A PLANE! IT'S A FOUNDER!

When first encountered in the *Deep Space Nine* series, Odo demonstrates his extraordinary power of shape-shifting. He is a Founder, a race that is fluid in its natural state. With exertion of energy, and through training and practice, Founders are able to assume the forms both of living and inanimate objects. According to Michael and Denise Okuda's authoritative *Star Trek Encyclopedia*, these imitations are so thorough that Founders exhibit all the properties of the imitated object, including the energy readings needed to avoid detection by scanning devices. Let's take a few readings of our own.

Odo's mass is initially portrayed as greater than one would expect to encounter in a humanoid of his size (*DS9*: "Vortex"). Well, he's made of liquid, and unless his constituent goo is significantly lighter or heavier than water, he probably weighs about the same as the rest of us, or about 180 pounds. Think about that, the next time you see Miles O'Brien carrying Odo around as a backpack.

Miles's back doesn't have it easy when Odo becomes a backpack. A shape-shifting organism still has to be made of the same number of atoms and molecules; Odo can change his shape, maybe even his volume (by stretching or collapsing himself), but he can't change his mass. When Odo takes the shape of a mouse, he might become as small as a mouse, but he'll still weigh 180 pounds.

What about shape-shifting itself? Might shape-shifters exist?

After all, amoebae change shape. So do schools of fish as they swim, and one could argue that a school of fish is "alive." It shows purposeful movement, acts to protect itself, consumes, and excretes. But amoebae and schools of fish maintain areas of coherent internal structure. They aren't homogenous matter, like Odo. We'll have to look for another shape-shifting life-form. Slime molds come as close as any earthly example to shape-shifting. These primitive creatures are colonies of single-celled organisms that coalesce into a moving mass. No one would ever mistake a lump of slime mold for a mouse, however. The gooey organism (or organisms, since slime molds are both a single entity and a colony) stays gooey, no matter what shape it takes. It cannot support its own weight if it is taller than a centimeter or two. In order to stand up, you need stiff supports—either bones or a shell—so a slime mold wouldn't be a good model for a sentient shape-shifter who wanted to impersonate a humanoid.

Similarly, for Odo to imitate the shape and surface properties of chairs, glasses, and spinning tops, he has to have the properties of those molecules. In many cases, he has to become solid. To accomplish the change in physical state from liquid to solid and back again would require energy going in and out. There would be some giveaway to the transformation—heat, light, or some other form of radiation would be emitted by a Founder that was maintaining a solid shape. The Founders would not be able to confuse tricorders.

The other miracle here is that Odo and the Founders are sentient. With a really active imagination, we can hypothesize a sentient organism that has some flexibility of shape, supposing that the neural network of the organism would still conduct signals. The Founders, however, are liquid. It is difficult to imagine a being that could be stirred and still maintain enough structural integrity to encode information. On the other hand, no one has ever stirred a Founder—at least, not in any episode we've ever seen.

Perhaps the key is that even while shape-shifting, some basic integrity of form is maintained. Founders might be like moldable Teflon webbing—capable of taking many forms but always with part A next to part B, and so on throughout the matrix. We know that when bits of Founders break off, the pieces revert to the gel state. This suggests that when the connection to the rest of the Founder is severed, the limb or whatever body part it might be does not receive enough neural signals to maintain a shape. We have also seen Odo blend into the Great Link and still retain his integrity as an individual. If Founders consist of a gelatinous neural net, they might be able to link with any other Founder who touched some part of their substance. Maybe they *are* like slime molds, after all.

What is still lacking, however, is any type of chemistry that could account for the ability to change state from liquid to solid and still code information. And speaking of codes, have you noticed that the genetic code seems to be broken now and then?

IT'S SO HARD TO EVOLVE

Worf was not himself. He'd spent a bad night sleeping on the floor and he had this awful, acid taste in his mouth. Maybe that's why he was pounding on the door of sickbay. Or maybe he knew that Deanna was turning into a frog and he thought if he could just kiss her, she'd turn back into a beautiful maiden and they could live happily ever after. . . .

In *TNG:* "Genesis," we are dealing with a DNA meltdown. It is a fact that only about 10 percent of the human genome actually codes for proteins. Large sequences of the genome are taken up with genetic control elements, telomeres, repeats, and other regulatory sequences. There are still huge portions of the genome unaccounted for. Some scientists speculate that on the way to becoming human beings, we have acquired extra DNA from everything in our ancestral path. Pseudogenes (referred to as "introns" in the episode) are thought to be inactivated copies of once useful genes. Well, then shouldn't it be possible to switch on that part of the genome and devolve?

No.

We already know what devolving tissue looks like: tissue that de-differentiates and begins to grow actively is cancer. Devolving humans would get lumpy and die. They would not turn into frogs, lemurs, or even cavemen. They most especially would not turn into spiders, as the unfortunate Lieutenant Barclay does when stricken with a bad case of intron-itis.

But let's play with this idea for a moment. If we were to evolve backward, we would not become gorillas or chimps, because we didn't evolve from them. Gorillas and chimps and we ourselves all evolved from a common ancestor. Nowhere in our evolutionary wanderings were we ever spiders (which are arthropods). Way, way back there, arthropods and protovertebrates shared a common ancestor. Now, if Lieutenant Barclay were to devolve into something really primitive and horrible, he'd be more accurate if he chose a sponge, but check the phylogenetic family tree. Pick your favorite primitive creature—you never know when you might get Barclay's Protomorphosis syndrome.

And while we're on the subject of devolution, if you are in the process of devolving, it's just no fair modifying dead tissues. Only the parts of your body that are actually alive and dividing can change when the genome produces a new protein. The living part of hair is the hair follicle, the cells at the base of the hair. These cells manufacture a complex protein product that we recognize as hair, but hair itself is dead. A typical part of the aging process, for most of us, is the loss of pigment cells from the hair follicle; when melanocytes are disabled, the hair turns white beginning at the roots, where the hair emerges from the follicle. The devolving

Riker would become a brunette Cro-Magnon with black roots. Which brings us to. . . .

IN SPACE, NO ONE STAYS BALD FOR LONG

Captain Picard and his alter ego, Patrick Stewart, have given baldness the respect it deserves. Inconsistently, however, no one else on the crew can stay bald.

If you lose your hair due to telekinetic aging, DNA resequencing, or for any other reason, you don't get it back until it grows back—that is, unless you serve on the *Enterprise*. Kirk, McCoy, Scotty, and even Spock turned gray in "The Deadly Years," but when their rapid aging was reversed, their hair returned to its former pigment and full texture. Dr. Pulaski in *TNG*: "Unnatural Selection" and Deanna Troi in *TNG*: "A Man of the People" also demonstrated this wonderful ability. But it's Geordi and his friend Susan (in "Identity Crisis") and nearly the entire crew of the *Enterprise*-D in "Genesis," who persuade us that hair growth in space is something special. Although their hair is completely lost when they turn into naked ultraviolet-light beings, it returns to its previous length, color, cut, and style when the DNA damage is fixed—apparently within a matter of hours or days. Of course, this peculiarity might be due to a virus. . . .

IF YOU CAN'T EXPLAIN IT,
IT MUST BE A VIRUS

When the crew gets really strange—Sulu turns into D'Artagnan (*TOS*: "The Naked Time"), or Tasha Yar makes love to Data (*TNG*: "The Naked Now"), or people take showers with their clothes on—you just know it's got to be a virus!

Viruses straddle the boundary between life and nonlife. It is important to understand that while both viruses and bacteria can be considered "germs," in that both are invisible to the eye and may cause disease, viruses and bacteria are very different. Bacteria have all the components of a real cell (except a nucleus and mitochondria—their DNA just floats loose in their cytoplasm). Viruses consist of a protein coat and just enough DNA to program their own structure and reproduction. Some scientists leave viruses outside the definition of *life-form* altogether, arguing that viruses are merely complex organizations of molecules which cannot move with intent, eat, breathe, develop, secrete, or excrete. There are many properties of viruses which we do not understand. We are quite certain, however, that they obey the law of gravity.

In the *Voyager* episode "Macrocosm," the crew is nearly wiped out by a virus that infects the biogel packs used as sophisticated circuitry relays and then infects engineer B'Elanna Torres when she attempts to make a repair. Once they have been exposed to Torres's "Klingon growth factor," the viruses grow to the size of gnats, then mosquitoes, then dragonflies, and finally they are the size of hovering pillows. They herd the crew into the mess hall and hold them as living incubators for baby viruses. Janeway does a Ripley (a.k.a. Sigourney Weaver) and retakes her ship with the help of an antibody bomb supplied by the Doctor.

The *Star Trek* writers have undoubtedly heard of airborne viruses, like the influenza virus. But airborne viruses stay aloft because they are tiny—on the order of a few microns. At this size, they have so little mass that gravity isn't sufficient to make them settle to earth when air currents keep them buoyant. Think of leaves blowing in the wind. Compared with viruses, leaves are massive, and when the air is still, leaves settle to earth. Viruses are just too tiny to settle. (So is cat dander, by the way, which is why putting Spot in the next room doesn't help Worf's allergy when he visits Data's quarters.)

Even macroviruses should obey the laws of physics. Viruses the size of pillows would need wings to fly, but then they wouldn't be viruses anymore. We are also quite sure that even the largest of viruses lacks the mental capacity to herd a crew together, hold them in a room, and attack Janeway, all the while making angry buzzing noises.

Viruses seem to hold a certain fascination for the *Star Trek* writers, as they do for scientists. Viruses can do many things, but another thing they can't do is code for memories. The physiology of memory remains a great mystery: the current best guess is that memories are coded as a specific pattern of simultaneous neural firing in the brain. Chemical storage theories for memory also exist. Both theories require complex networks of neurons to be simultaneously engaged. But viral particles are so tiny that thousands could fit inside one single neuron. Since viruses cannot code for memories, people can't "catch" memories (or Post-Traumatic Stress disorder, either, for that matter) as if they were contagious diseases—the way Tuvok "catches" the memory of his dying crewmate Dmitri on the *Excelsior*, in *VGR*: "Flashback." To postulate that the virus "jumps ship" as one host is dying and transfers itself to a new host is to give way too much credit to viruses.

One more observation about viruses in the twenty-fourth century: then, as now, measurements are a key part of scientific observation. Accordingly, it is very important to be able to describe size accurately. We are quite certain that atoms in the twenty-fourth century will still be smaller than molecules, that molecules like proteins and DNA will be smaller than viruses, and that viruses will be smaller than bacteria. This means that there will be no "subatomic bacteria," as there are in *TNG*: "A Matter of Honor." With a large stretch of imagination, we can allow that there might be deep-space anaerobic bacteria capable of corroding a starship's hull, but we are sure that they won't be subatomic in size. Neither is anyone likely to observe a molecule of DNA with a bar code stamped on the chemical bonds of the base pairs, as in *VGR*: "Scientific Method."

WHERE HAVE ALL THE FLOWERS GONE?

Ever notice how most Class-M planets the crew explores look just like Southern California? The Panspermia hypothesis seems to be right at least as far as vegetation is concerned. With the exception of exo-exotics, like Sulu's moving pet orchid or the humming flowers on the world of the Talosians, most plants in the *Star Trek* universe are garden-variety scrub grasses and potted ferns.

And while we're talking about plants, have you noticed how many Class-M planets don't have any at all? Maybe the crew just happens to pick landing sites that are rocky and barren. But without plants, you don't have oxygen, and without oxygen you don't have a Class-M planet. Of course, those rocky planets are the best places for finding caves that collapse on anyone wearing a Starfleet uniform.

THAT'S NOT HOW THE BRAIN WORKS

Uhura is humming, and Nomad, the neurotic robot of *TOS*: "The Changeling," hears her. Nomad becomes interested and floats over to ask her what she's doing. "I was singing," Uhura replies. So Nomad asks Uhura to think about music, and then he scans her brain—but he empties it, too.

Here the *Star Trek* writers have confused two different philosophical constructs. When we consume food, we digest and destroy it. When we consume information, we download (or upload, for you optimists) sensations, facts, and feelings into our brain memory banks—but the information isn't destroyed in the process. As you read this book, the words (we hope) are staying on the page, so that your friends and neighbors can read it, too (we hope).

The fascinating thing is that there is indeed a way in which Nomad could scan Uhura's brain and perhaps learn about music. Thoughts are represented by patterns of neuronal firing. These patterns flicker and fade rapidly, but by using functional brain imaging techniques and taking pictures of the brain during concentration tasks, neuropsychiatrists are beginning to understand how the brain codes certain language symbols. If Nomad had scanned 1,000 or so humans while they were thinking of specific topics, it might have learned how to read brain-firing patterns well enough to obtain a concept of music from Uhura. But Nomad would not be emptying her brain in the process.

There is an even bigger blooper in "The Changeling." One of the final scenes shows Nurse Chapel tutoring the now empty-headed Uhura in reading, using an elementary primer. McCoy reassures Kirk that Uhura will be "back at her post" in a couple of weeks or so. Gee, and she spent four years at Starfleet Academy learning it the first time. Wish we knew how to study that fast!

THE SHUTTLE BAY DOORS ARE OPEN

Geordi La Forge and Beverly Crusher are trapped in the shuttle bay when the *Enterprise*-D encounters a serious "spatial disturbance" that causes widespread damage throughout the ship. The bad part is, they are surrounded by radioactive cargo. The really bad part is, there is a "plasma fire" in the wall of the shuttle bay. The really, really bad part is, the plasma fire might make the radioactive cargo blow up. What would you do?

Of course! You open the shuttle bay doors just wide enough to let the vacuum of space sweep the cargo out and the lack of oxygen put out the fire. Geordi tells Beverly to "Hang on tight!" Indeed. How long can you hold your breath? It's only a near-absolute vacuum, and it's only for a minute. Who needs eardrums anyway?

Worf also had a problem with vacuum during his space walk battle with the Borg in the movie *First Contact*. His pressure suit developed a leak. No problem for our hero. He fashions a tourniquet out of some handy tubing (with a Borg hand attached, but sometimes you have to make do). Let's hope he tied it *really* tight, because there's that vacuum-of-space thing. All the blood in his body would end up in his foot, which would probably explode under the stress. Biology can be messy!

WHATEVER HAPPENED TO THE KLINGONS?

This particular blooper is so bad that it's hard to discuss, which is how the *Star Trek* writers and the Klingons have both decided to handle it. You'll recall that when Kirk first encounters the Klingons, they look like Genghis Khan come back from the dead—scary, with dark mustaches and slanting eyebrows, but they are no bigger than the average Earthling, and they have uncorrugated foreheads. Worf would have considered them wimps. But by the time *The Next Generation* and *Star Trek III: The Search for Spock* appear, Klingons are several inches taller, broad-shouldered, fanged, and sporting their brutal forehead ridges. As far as we can tell, when the *Star Trek* movie budget grew, it caused tremendous evolutionary changes among the Klingons. When the crew of Deep Space 9 traveled back in time, they were astonished at the appearance of the original Klingons. Worf merely grunted, "We don't like to talk about it." (*DS9*: "Trials and Tribble-ations")

We have come up with a theory, however. We think there was a little nerdling Klingon kid who got tired of always getting beaten up in Batlith class. His parents were concerned. His father tried coaching the local Warrior Scout troop. His mother hired a private martial arts tutor. No good. The kid just kept getting beaten up. Finally, he decided that the only way he'd ever improve his ranking in Batlith Little League was to get bigger and stronger. So he worked hard on his science-fair genetics project that year. Just before it was ready to go, his little sister sneezed near the culture dish and introduced a bunch of wild viruses into his vector. When he injected himself with the vector carrying the recombinant bigger-stronger gene, it turned out to be wildly contagious. The fashion industry on Qo'noS took years to recover.

We shall leave you here, friend reader. We hope we have completed our mission objectives: to entertain, to teach, and to share some favorite *Star Trek* moments. We have only touched on the material contained in the 79 original series episodes, the 178 *Next Generation* episodes, and the 7 movies. The crews of Deep Space 9 and *Voyager* are still on active duty, and we understand that Picard hasn't retired from Starfleet yet. There is a lot of universe still to explore. Back here on twentieth-century Earth, there is certainly more biology to explore. The Human Genome Project and the breakthroughs in neuropsychiatry are happening right now. The twenty-first century promises to be a great one.

Live long and prosper!

Qapla'!

GLOSSARY

amino acid carbon molecule containing a nitrogen group and a carboxylic acid group; amino acids link (in chains of less than 10 to several thousands) to form **proteins.**

amoeba a single-celled nucleated organism, usually found in pond water, with a flexible outer membrane.

antibodies proteins that mark foreign particles or damaged cells for attack by the body's **immune system.**

asexual reproduction producing offspring by budding, cloning, cell division, or other means that does not combine genetic material from two individuals of the species. See **sexual reproduction.**

bacterium a single-celled organism without a nucleus, with rigid polysaccharide cell walls and a single circular **chromosome** in the **cytoplasm.** See **prokaryote.**

base pair the primary unit of measurement in DNA, composed of a purine (adenine or guanine) and a pyrimidine (thymine or cytosine) bonded to each other. Adenine (A) pairs with thymine (T), and guanine (G) pairs with cytosine (C). The normal human **genome** has 6 billion base pairs.

catalyst an agent, such as an enzyme, that facilitates a chemical reaction.

cell differentiation the developmental process by which young cells change into mature and more specialized cells.

chimera in mythology, an animal made of parts from several other animals; for example, a griffin has the head, claws, and wings of an eagle, and the body, hind legs, and tail of a lion. In molecular biology, a chimera is an organism that has been genetically engineered by combining the genetic material of two individuals, which may or may not be of the same species.

chromosome a single polymeric molecule of DNA in cells; an average chromosome contains about 125 million **base pairs**, which comprise 40,000 to 100,000 genes. The human genome is made up of 23 pairs of chromosomes (for a total of 46). Normal women have two X chromosomes as the 23rd pair; normal men have an X and a Y chromosome as the 23rd pair.

clone 1. an offspring produced by **asexual reproduction** (or genetic engineering) and genetically identical to the parent; 2. a genetically identical sibling.

commensalism a relationship between two organisms in which one clearly benefits and the other is not harmed; for example, buffalo birds that follow bison and eat insects stirred up by the grazing animal's movements have a commensal relationship with the buffalo.

cytoplasm the components of a cell that are not part of the nucleus.

deoxyribonucleic acid (DNA) the biochemical molecule that makes up genes; a double-stranded helical polymer with each strand composed of pyrimidine or purine bases connected to repeating phosphate and sugar (deoxyribose) backbones; the double helix is held together by the pairing of bases.

embryo an organism in early development, following fertilization of the egg by the sperm and prior to the formation of the major organs.

enzyme a catalyst that helps biological chemical processes to occur more rapidly and use less energy. Enzymes in plants and animals are made of proteins and are designated by the suffix *-ase,* as in lactase, which assists in breaking down lactose, a simple sugar.

eukaryote an organism made up of one or more cells, each of which has a nucleus containing DNA, as well as extranuclear cell bodies (such as **mitochondria**) that contribute to cell function. All multicellular animals and all plants except the blue-green algae are eukaryotes. See **prokaryote**.

fetus the developing organism following organ formation.

free radicals chemicals with highly active electrons that are capable of disrupting the chemical bonds of other molecules; in living systems, free radicals can damage proteins and nucleic acids and other complex molecules.

gamete egg cells and/or sperm cells; the DNA complement of gametes is half the amount in a mature cell of the organism (one **chromosome** of each pair). One egg cell and one sperm cell must combine in sexual reproduction.

gene the amount of DNA that codes for a single **protein**. The human **genome** contains about 150,000 genes.

genome the entire genetic code of an organism, as contained within individual cells.

genotype the genome; for example, men and women are of the same species but have slightly different genotypes because of the X and Y chromosomes.

gradualism a theory of evolution that emphasizes the slow accumulation of small changes in the structure of an organism which gradually results in the creation of a new species over a large number of generations; such changes are selected by changes in the environment. See **punctuated equilibrium**.

Hayflick limit the maximum number of times a cell is able to divide. For human cells, the Hayflick limit is 50 to 70 generations. See **telomere**.

homeobox genes (Hox genes) a family of genes that help determine the position of body segments in higher organisms.

homeostasis the tendency of organisms to seek and maintain a stable, balanced state, as (for example) body temperature.

homologous recombination a genetic engineering technique whereby a copy of an altered DNA sequence is inserted into its correct location in the genome and replaces the naturally occurring sequence.

homology 1. similarity of structure or function (for example, the wings of bats and birds are said to be homologous limbs for flight); 2. in molecular biology, the similarity in sequences of base pairs between two or more segments of DNA.

hormone a bioactive molecule that is produced in one part of the body and circulates to its target tissue in another part of the body; adrenaline is an example.

Human Genome Project multi-institutional and multinational coordinated research effort to determine the base sequence of the entire human genome and map the location of all genes to their respective chromosomes.

humanoid an intelligent being that has the same body plan as humans.

hybrid the offspring of an interspecies mating. Mules are terrestrial examples; Spock and B'Elanna Torres are *Star Trek* examples.

hydrolysis chemical process by which water is removed from a molecule or combination of molecules.

immune system the body system that defends against foreign material and germs; **antibodies**, the complement system, and white blood cells are the principal components of the human immune system.

imprinting the genetic process by which genes from the male and female parent are differentially repressed or expressed.

knockout mouse a mouse or breed of mouse developed for research purposes which is lacking one or more specific genes.

metabolism the sum of the chemical and physiological activity of the body that allows it to use matter to produce energy and sustain its activities.

mitochondrion (pl., mitochondria) an extranuclear structure in the cells of **eukaryotes** which uses the oxygen from respiration to burn food and generate energy.

morphology the structure of an organism; refers not simply to anatomy but also to the developmental structures that preceded the current state. For example, the anatomy of a building refers to its walls, floors, measurements, and material. The morphology refers to how these function together to make the building useful and also to the various stages of the building's construction.

morula a developmental stage of the **embryo,** at which several cell divisions (but no growth) have occurred; it is a solid ball of from 16 to 64 cells, which is usually the size of the original fertilized egg cell.

mutation a change in a gene; a point mutation is a change in a single **base pair** of the gene.

mutualism a relationship between two organisms which is beneficial to each.

neuron, or nerve cell the nerve cells in the brain and spinal column responsible for processing the electrochemical signals of the central nervous system; there are 100 billion neurons in the human brain.

neurotransmitter a naturally occurring molecule that conveys a signal from one **neuron** to another.

nucleic acids DNA and RNA.

organic compounds naturally occurring carbon-based molecules found in life-forms.

Panspermia hypothesis the idea that life arose on Earth from primitive extraterrestrial microorganisms: originated with the 19th-century physicist Svante Arrhenius, elaborated in the 1970s by biologists Francis Crick and Leslie Orgel.

parasite; parasitism a relationship in which one organism is dependent upon another for some or all of its vital needs—food, shelter, reproduction—and is harmful to its host; a tapeworm is an example of a parasite.

phenotype the product of the genome; usually refers to an organism's physical form, coded for by the organism's DNA.

photosynthesis a metabolic process in plants by which sunlight is used to fix carbon dioxide from the atmosphere and manufacture glucose, with oxygen released as a byproduct.

polymorphism a region of the genome common to the general population but containing base-pair sequences that vary from individual to individual, which can therefore be used to identify an individual's DNA. There are several hundred thousand polymorphisms in the human genome.

prion a disease-causing **protein** with an atypical folded structure.

prokaryote a single-celled organism lacking a nucleus (chromosomal material is scattered throughout the **cytoplasm**) and usually having a rigid or semirigid cell wall; **bacteria** and blue-green algae are examples of prokaryotic organisms. See **eukaryote**.

protein a complex biochemical molecule composed of a chain of amino acids; proteins make up the structure of most animal tissues—muscle, skin, tendons, etc.

punctuated equilibrium a theory of evolution that emphasizes the rapid accumulation of large changes in the structure of an organism resulting in **speciation** after a small number of generations; such changes may be selected for by catastrophic alterations in the environment; prolific **speciation** is followed by long periods of stability or minor evolutionary change.

ribonucleic acid (RNA) a single-stranded nucleic-acid polymer composed of purine and pyrimidine bases connected to a repeating phosphate and sugar (ribose) backbone. Messenger RNA (mRNA) is the product of gene **transcription** in the cell nucleus and conveys the genetic blueprint to the protein-manufacturing site in the cell cytoplasm; transfer RNA (tRNA) assists in protein manufacture in the cell cytoplasm by translating the mRNA to an amino acid sequence.

sequence *n.* a portion of DNA containing a specific order of base pairs; *v.* the process of analyzing and determining the order of the base pairs comprising a segment of DNA.

sexual reproduction producing offspring by combining genetic material with another individual of the same species.

speciation the process of forming multiple species from a single ancestor.

stem cell a cell with potential to become one or more specialized mature tissue cell types; embryos are mainly composed of stem cells; in the mature organism, stem cells are found in the bone marrow, gastric mucosa, and gonads.

symbiont an organism that is dependent upon another organism for its food, environment, and/or other vital needs.

symbiosis any of several relationships between organisms, including mutualism, commensalism, parasitism, and less interdependent relationships such as predator and prey or master and pet.

synapse the place where two **neurons** connect.

telomere a special repeating sequence of nucleic acid at the end of a **chromosome**, which does not code for a gene product; the loss of teleomeres is thought to be important in causing cells to stop reproducing after a certain number of cell generations. See **Hayflick limit.**

transcription the process by which DNA is read into RNA and the genetic instructions are taken out of the cell nucleus to the cytoplasm. A related process is translation, in which the RNA is read and the genetic blueprint directs construction of a protein—this process occurs in the cell cytoplasm. See **ribonucleic acid.**

vector in molecular biology, an organism or fragment of an organism used to insert DNA or RNA into a living cell; viruses and plasmids are commonly used vectors.

virus the simplest form of life on Earth, consisting of nucleic acids wrapped in a protein coat. Viruses do not consume or excrete and are not capable of voluntary movement. For these reasons, some scientists dispute their qualifications as a form of life.

SELECTED READINGS AND REFERENCES

STAR TREK REFERENCES

Krauss, Lawrence M., *The Physics of Star Trek* (New York: Basic Books, 1995).

Nemecek, Larry, *The Star Trek The Next Generation Companion* (New York: Pocket Books, 1995).

Okuda, Michael, and Denise Okuda, *Star Trek Chronology: The History of the Future* (New York: Pocket Books, 1996).

Okuda, Michael, Denise Okuda, and Debbie Mirek, *The Star Trek Encyclopedia: A Reference Guide to the Future*, updated and expanded edition (New York: Pocket Books, 1997).

Trimble, Bjo, *Star Trek Concordance: The A to Z Guide to the Classic Original Television Series and Films* (New York: Citadel Press, 1995).

Sternbach, Rick, and Michael Okuda, *Star Trek The Next Generation: Technical Manual* (New York: Pocket Books, 1991).

OTHER REFERENCES AND RECOMMENDED READINGS

Alberts, Bruce, and Dennis Bray, Julian Lewis, Martin Raff, Keith Roberts, and James D. Watson, *The Molecular Biology of the Cell*, 2nd edition (New York: Garland, 1989).

Austad, Steven N., *Why We Age: What Science is Discovering About the Body's Journey Through Life* (New York: John Wiley & Sons, 1997).

Carlisle, David Brez, *Dinosaurs, Diamonds, and Things from Outer Space: The Great Extinction* (Stanford, Cal.: Stanford University Press, 1995).

Diamond, Jared M., *The Third Chimpanzee: The Evolution and Future of the Human Animal* (New York: HarperCollins, 1992).

Gerhart, John, and Marc Kirschner, *Cells, Embryos, and Evolution: Toward a Cellular and Developmental Understanding of Phenotypic Variation and Evolutionary Adaptability* (Oxford, U.K.: Blackwell Science, 1997).

Gould, James L., and Carol Grant Gould, *Sexual Selection* (New York: Scientific American Library, 1989).

Gould, Stephen Jay, *The Panda's Thumb: More Reflections in Natural History* (New York: W. W. Norton, 1980).

Holliday, Robin, *Understanding Aging* (New York: Cambridge University Press, 1995).

Hubel, David H., *Eye, Brain, and Vision* (New York: Scientific American Library, 1988).

Landau, Terry, *About Faces, The Evolution of the Human Face: What It Reveals, How It Deceives, What It Represents, and Why It Mirrors the Mind* (New York: Doubleday, 1989).

Mallowe, Eugene, and Gregory Matloff, *The Starflight Handbook: A Pioneer's Guide to Interstellar Travel* (New York: John Wiley & Sons, 1989).

Mange, Arthur P., and Elaine Johansen Mange, *Genetics: Human Aspects* 2nd edition (Sunderland, Mass.: Sinauer Associates, 1990).

Norman, David, *Dinosaur!* (New York: Prentice Hall, 1991).

Sagan, Carl, *The Demon-Haunted World: Science as a Candle in the Dark* (New York: Random House, 1996).

Shapiro, Robert, *The Human Blueprint: The Race to Unlock the Secrets of Our Genetic Script* (New York: St. Martin's Press, 1991).

Strahler, Arthur N., *Science and Earth History: The Evolution/Creation Controversy* (Buffalo: Prometheus Books, 1987).

ADDITIONAL SOURCES

For interested Trekkers who want to stay at the forefront of science (biological or otherwise), several general journals are available. They include *Science News, Scientific American, Natural History,* and *The American Scientist.* Besides being accessible to the scientist and nonscientist alike, current issues nearly always have an article or two that relates to scientific issues relevant to *Star Trek* (if you read between the lines). We used them frequently in preparing this book. In addition, two weekly interdisciplinary journals for professional scientists, *Nature* and *Science,* frequently publish brief summaries of current outstanding research. These summaries are usually in less technical language than the scientific papers in these journals, and are often accessible to general readers. If you really want to find out what scientists are doing, these are two important journals to read.